Mental

Math

A Quick and Easy Guide to Mental
Math and Faster Calculation

*(Enhance Your Math Skills and Solve Everyday
Math Problems Easily)*

Robert Potter

Published By **John Kembrey**

Robert Potter

Mental Math: A Quick and Easy Guide to Mental Math and Faster Calculation (Enhance Your Math Skills and Solve Everyday Math Problems Easily)

ISBN   978-1-998769-76-6

Table Of Contents

Chapter 1: Multiplying Big Numbers Quickly

Mental math skills are extremely important ones to have for use on a regular day to day basis but they require long hours of diligent practice to maintain and improve. Once basic proficiency in math has been established, it is never enough to just halt your progress wherever you decide to. Learning skills in math is just like learning skills in any other field. It is a cumulative process that requires lots of time and dedication to be done effectively.

There are many different facets of mathematics that you have to learn in order to gain any proficiency in the subject. Throughout the course of this book, we will go over some of the most important of these facets especially through the optic of developing mental math skills.

We should start with a basic aspect of math, multiplication. It takes a number of different strategies to develop mental calculation abilities in regards to multiplication. Once all of these strategies are learned, they can easily be applied to specific situations that apply to each respective strategy.

When you think of most of the math that the average person finds him or herself doing on a day to day basis, it is usually not done by means of calculators and even more rarely done by means of written computation. It is almost always done instead by mental computations. With this being said, it is no wonder why skills developed in mental math computations are always incredibly useful ones to have. Sadly though, many (if not most) neglect to develop their mental math skills beyond a certain age and are put at a disadvantage for doing so. As far as students are concerned, developing mental math skills can also help them to get a better understanding of both number sense and number properties.

Multiplication by single digit numbers
The first type of mental multiplication that we should now go over is multiplication by single digit numbers. At first, students can use this strategy to multiply smaller numbers. Once they have developed skills with these they

can then move on to bigger numbers. For example, a student can multiply 3 four times to get 12. If he or she then multiply 3 by 12 then the student gets 36.

Multiplication by powers of ten
The next step in multiplication is multiplying by powers of ten. This is much easier to do as it only results in another decimal point being included in the final answer. To multiply by powers of ten, all you need to do is multiply the other number by the first number in the power of ten. For example, four times ten would come out to 40, while four times one would come out to four. As you can see, the decimal points, or zeros, in these cases just carry over with the rest of what is being multiplied.

If, however, the numbers do not end in zero, then the first decimal point listed should have a zero put behind it when multiplying. For example, 15 times 6 would translate to 5 times 6 (30) and 10 times 6 (60). 30 plus 60 would then equal 90, so 15 times 6 would be 90.

One strategy that students often use is the visualization of written algorithms. This is not an advisable strategy to use because it makes making computations much less efficient and much more difficult. Algorithms only work well when written down on paper. There are other much more useful methods for multiplying mentally.

Mental estimation

Mental estimation is another very important skill for a student to learn. This is especially useful when answers are specific to an extent at which it would require a calculator to find them. For example, estimation would have to be used when finding out the area of a yard on the fly. If the fencing was 20ft. by 10ft., the area would be 200ft. A student could not, however, know definitively the length of the fence, so he or she would have to estimate in this case. To better estimate mentally you must know how to do mental calculations, round to convenient numbers, and be able to determine whether or not an answer is too big or too small.

Again, it requires frequently repeated practices to develop skills in making mental math calculations. There are some very helpful activities for developing these skills which should now be mentioned.

Activities

Activity 1
Use single digit strategies to improve on your most basic multiplications. Once the more basic single digit strategies have been gone over for some time, you can then move into bigger numbers and more complex calculations, such as those by factors of ten.

Activity 2
Apply the distributive property to the single digit strategies previously gone over. The distributive property can be solved in these three steps:

1.   Multiply any term placed outside of the parentheses by each and every term located within the parentheses. In doing this you will distribute the outside term among the inside terms.
2.   Combine like terms. If any terms in the equation are the same, combine them into one term.

3.   Add the terms left to solve the equation.

Let's look at an example of solving an equation with the distributive property:
$5(2+8) = 5(2) + 5(8) = 10+40 = 50$, so, $5(2+8) = 50$

Activity 3
Split larger numbers into factors that will fit into them and then solve the equation using these numbers. We will explore this further later.

Activity 4
Keep a mental track of the equations that you are solving throughout activities one and two. This will engage the circuitry involved in solving these problems long term and you will commit these strategies to your long-term memory.

Activity 5
Keep an eye out when multiplying large numbers for factors that allow for easy multiplication. For example, 8 times 21 can be distributed as 8 times 20 (160) plus 8 times 1 (8), so your answer here would be 168.

Now that our basic activities have been outlined, we should delve deeper into how to go about these activities in the easiest ways starting with activity number 1: multiplying by single digits.

To multiply by 2, take the number being multiplied and double it. To multiply by three, double the number and then add to that the number once more. To multiply by four, double the number two times over. To multiply by 5, multiply the number by 10, and then halve the answer. There are two main ways to go about multiplying by 6. Either you could multiply the number by 3 and then double that number, or you could multiply the number by 5 and then add the number once more. There is no easy way to multiply by 7. You must just count up by 7s by whatever you are multiplying it by. To multiply by 8, double the number three times over. To multiply by 9, first, multiply the number by 10 and then subtract the number from that answer. To multiply by 10, just add a zero to the number (but watch out for decimal places carefully).

We can then extend all those principles laid out for single digit numbers to use for dealing with larger numbers. To use the number 100 for example, there are many ways to multiply easily by 100 and numbers around it.

Multiplying by numbers around 100
If you wanted to multiply by 101, you could multiply by 100 and just add the number that you are multiplying once more. To multiply by a 102, first, multiply by 100 and then add the number times 2. To multiply by 99, first, multiply the number by 100 and then subtract the number being multiplied from that once. To multiply by 98, first, multiply by 100 and then subtract the number being multiplied times 2. As you can probably see by now, many of the same rules mentioned in the first activity apply just on a grander scale.

The next helpful thing that we can do is to split the factors that we come across in equations with bigger numbers into smaller factors that can come out of them. For example, 20 times 36 is an equation with very large numbers, but we can split these numbers into multiplications that will add up to them. 20 could here become 5 times 4, and

36 here could become 6 times 6. So the result here would be 5 times 4 times 6 times 6, equaling 720, so 20 times 36 would equal 720. Another example would be 16 times 18. 16 could become 4 times 4, and 18 could become 6 times 3. So we would eventually get 4 times 4 times 6 times 3. This would equal 288, so 16 times 18 would equal 288.

Next, we should make it a point to keep track of the numbers in any given equation that you are solving, especially 0s. Multiplying by numbers which end in 0 is usually a very easy thing to do but the only stipulation is that you have to keep track of the 0s in these cases. On example of doing this could be in solving 30 times 40. This could be understood as 30 groups of 40. 30 groups of 40 could then been seen as 10 groups of three times 40. 3 groups of 40 would then be 120, and 10 groups of 120 would then be 1,200, so your final answer would be 1,200. As you can probably see, multiplying using this technique is simply not possible without paying special attention to keeping track of the 0s.

There should also be mentioned tips on multiplying by other special numbers between 1 and 100. These, like the ones near 1 and the ones near 100, can be multiplied more easily when you take into account the relations to other numbers.

Some examples of these numbers are listed below:

- To multiply a number by 50, multiply it by 100 and then halve that number.
- To multiply a number by 25, multiply it by 100 and then divide that answer by 4.
- To multiply a number by 500, multiply it by 1,000 and then divide the answer by 2.
- To multiply a number by 250, multiply it by 1,000 and then divide the answer by 4.
- To multiply a number by 125, multiply it by 1,000 and then divide the answer by 8.

Again, these tips and activities should be used and applied on a day to day basis to achieve any significant results. Mental math is just like anything else in that you need to be persistent to develop skills in it. If you do, however, stick to using the tips laid out here daily for some time, and put genuine effort into getting better at making these calculations, then you should eventually see some improvements in your performance and ability.

Proportions and Ratios

The next area which would be helpful to go over is proportions and ratios. We should start off by first defining these two terms. A ratio is essentially a means of comparing two quantities, whatever they might be made to represent. An example of a ratio would be miles per hour. Here, we are comparing the number of miles with the number of hours that it takes to drive those miles. A proportion, on the other hand, is an equation which assumes that the values of two or more different ratios are equivalent to one another. If the two equal one another, then we call the two being "in proportion" with each other.

We should now go a little more in-depth with our discussion of each of these. A ratio, as mentioned above, is a comparison of two terms. These are usually divided by one another when ratios are expressed. Some examples of this are as follows:

x to y, x:y, x/y

Proportions

A proportion is another equation. The purpose of this one is to find what, if any, equations are equivalent to another. These can be extremely useful in everyday scenarios, for example, let's say that you are making cookies and the packaging says that one pack of cookie dough will make 20 cookies. A proportion that you could deduce from this is that two packs of the cookie dough would then make 40 cookies. If we were to write this out mathematically it would look something like this:

$20/1=40/2$

The writing out of a proportion requires more variables than does the writing out of a ratio because there are more terms in proportions. A proportion written out would usually be along the lines of "x is to y as z is to w", or, $x/y=z/w$.

Often times you will get ratios and or proportions in which some of the individual variables will be unknown to you. This is where some basic algebra skills will come into play as you will have to then solve these equations. If you are reading this as a non-math person then do not be intimidated. This is basic equations to follow and can be done easily by someone who does not even have a math background. Let's now look at another basic cooking example below:

You have to make 20 pancakes for a get together with family and friends. You know that making these twenty will require you to use 2 eggs. And so, 2 eggs will make a total of 20 pancakes. Using this information, how many eggs will it require to make 100 pancakes?

So, 20 pancakes=2 eggs, 100 pancakes=x eggs.

There are two common equations that could be drawn up here, but only one is correct:

Eggs/pancakes = eggs/pancakes
or
pancakes/eggs = pancakes/eggs

The correct equation, in this case, would be:
eggs/pancakes=eggs/pancakes

because this one puts out mystery number (x eggs) in the numerator, making it possible for us to solve for x.

Here is what we would do next:

1.   Write it out as x/100=2/20
2.   Then, we would multiply both sides by 100:
3.   100 times x/100, and 100 times 2/20, this would then become x=200/20, and finally, x=10.

It is, however, possible to solve for x even if the variable is in the denominator. This would just require us to use another method known as the cross product method. This one is not as commonly used but it could be argued that it is even easier than the method previously mentioned. The cross product, in this case, is the product of the first numerator of the proportion and the second denominator of it. The inverse is true for the second numerator; you would multiply it by the first denominator. Here is an example of this method using the cookie dough equation mentioned above. Again x, or 2, is represented on the bottom here:

20/1=40/2

So cross multiplying would look something like this: 1x40=2x20=40.

We should now consider again our basic equation for proportions:
$x/y = z/w$, to cross multiply this we would come up with something like this: $xy = zy$.

Scaling examples
The next topic that we should go over is scaling. This of the map keys that you have seen. They usually will show you some minuscule length like an inch and tell you that the length on the map is equal to another length, in reality, 100 miles for example. This is what is known as scaling. We often use this method to depict various sizes of things. Scaling usually consists of maintaining the proportions of a certain object, while adjusting its size in all sorts of ways. It remains the same shape, just not the same size. In scaling, you can either go up (enlarge) or go down (reduce) in the size of an object. Scaling factors are often represented by ratios, for example, 1:4 represents a quarter,

1:8 represents an eighth and so on. Thus, 1:4 of an object, in this case, would not be a quarter of the object but the object brought down to a quarter of its original size. When solving equations for the scaling of objects, it is necessary to use ratios to represent the scales. For example, if we were to reproduce a 1:4 size version of a 20ft wall, the calculation that we would have to use would look something like this:

$20 \times 1:4 = 20 \times \frac{1}{4} = 5$

If we do not know what we are scaling to, then x becomes our variable. For example, in a scale model of 1: x, x is our constant term, and therefore all the measures we could create would have to be 1: x- of the real measurement. The only thing left for us to do in this case would be to solve for x, which was mentioned how to do previously. The same would hold true when enlarging a figure. In depicting something in the scale of 2:1, for instance, all models to come would be twice as large as the original.

So let's move on to some word problems now. Say, for example, you are building a fence. 20 boards, in this case, could be used to build 5 feet-fence. With that being said, we now need to find out how many boards it would require to build 20 feet-fence. Let's now plug in the variables into the equation listed above for proportions, x/y=z/w.

What we are trying to solve for in this case would be z, and the equation would look something like this: 20/5 = z/20. Here, as you can see, our unknown variable is in the numerator. Our next step would look something like this:

20 x 20 = 5z, so, 400 = 5z, 400/5 = 80, and therefore, z = 80

Another example in which there is only one variable to solve for is 2:x = 3:9.

The first step that we would take here would be to convert the colons into slashes to put the equation in terms of two fractions: 2/x = 3/9

The next that we would need to take here would be to cross multiply our fractions.

This would look something like this: 2 x 9 = 3x

Now, the value of 3x is equal to the value of 2 x 9 so we would have to figure out the value of the first multiplication here, 2 x 9. 2 x 9 = 18.

And so its 18 = 3x here.

Next, to solve for x we would need to divide 18 by 3. 18/3 = 6.

And finally, x = 6.

Up until this point all of the examples that we have gone over have been pretty basic, but often times you will find yourself having to solve more complicated problems with proportions, even in your everyday life. We should now up our game with problems that are a little more complicated so that we will get a better view of just how much you can do when solving for proportions and ratios.

The next example is:

$(2x + 1):2 = (x + 2):5$

You have probably noticed more parentheses and variables here than usual. Do not be alarmed if you are not a math person though. Equations like this are easier to solve than they may appear to be for someone who does not go over algebra regularly.

The first step here, like the first step in the previously mentioned problem, would be to convert the colons to fractions. This would make the equation look something like this:

$2 x + 1 / 2 = x + 2 / 5$

Next, you would have to do a multiplication by the cross (keep in mind that you cross multiply the numerators with the denominators, not any terms within the parenthesis). This would make the equation then look like this:

$5 (2 x + 1) = 2 (x + 2)$

And so, after multiplying the inside terms with the outside terms, on either side of this equation, what we are left with looks something like this:

10 x + 5 = 2 x + 4

The next step is where this gets a little tricky. Here we need to combine like terms. When this is done we need to subtract 2 from 10 and 4 from 5 in this case. The result would look like this:

8 x = 1

And finally, we divide 1 by 8 to get ⅛, so x = ⅛

We should now look at a word problem. Here is our next example: 12 inches is exactly 30.48 centimeters, so, with that said, how many centimeters would there be in 30 inches? In this equation, the letter c will be used to represent centimeters. Our basic equation would then look something like this:

12 / 30.48 = 30 / c

Next, we need to cross multiply:

30 x 30.48 = 12 c

So, 914.4 = 12 c, and then we would divide 914.4 by 12: 914.4 / 12 = 76.2

c = 76.2

Next is another word problem. Word problems in these cases are especially useful because they give us an idea of what it is like to encounter these problems in everyday life and they prepare us for real-world problem-solving. Here is our next example: You have a metal bar that has a length of 10 feet and a weight of 128 pounds. With that being said, what would be the weight of a bar with the same density but a length of only 2 feet and 4 inches?

This is a more complicated problem than anything that we have yet come across. The first problem that we need to solve in this case is that of inch conversion, we can do this by finding the fractional form of the inches in relation to feet. Our equation for 2 feet 4 inches would look something like this:
2 ft. + 4 in. = 2 ft. + 1 / 3 ft. = 7 / 3 ft.

Now, to add all of the other variables:

10 / 128 = (7/3) / w

Next, we would have to cross multiply:

(7/3) x 128 = 10 w

298. 1 / 3 = 10 w

Next we would have to divide 298. 1 / 3 by 10:

298. 1 / 3 / 10 = 29. 8 1 / 3

So, in this case, the 2 feet 4-inch bar would weigh 29.8 ⅓

The next word problem that we will now move onto involves taxes, which makes it and its jargon a lot more useful to normal people than most of these other ones.

A property with an assigned value of $70,000 has a tax of $1,100. Meanwhile, there is another property within the same tax district as the original one. The tax imposed on this other property is $1,400. With this being mentioned, what is the assigned value of the second property mentioned here?

Here we are given two different categories to access: the assigned values of the property and their taxes. We should now put these into ratios to get our basic equation:

70,000 / 1,100 = v / 1,400

The next step then would be to cross multiply. Our equation now looks something like this:

70,000 x 1,400 = v 1,100

Next, 98,000,000 = v 1,100

So, now we have to divide 98,000,000 by 1,100:

98,000,000 / 1,100 = 89,091

So, v here equals 89,091, and therefore the assigned value of the second property would be $89,091.

As you can plainly tell by now, ratios and proportions are extremely important to know how to multiply. This skill may even come in handy more often than does basic, single digit multiplication at times. If you stick to the principles laid out here, when solving for ratios and proportions, you will be able to get better with these problems and their equations.

Chapter 2: Trigonometry And Its Uses In Real

Life

Trigonometry is a subject that often gets a reputation as being harder than it truly is. To start off our discussion on trigonometry, this is the study of the calculations involving triangles in real life (hence the "trig" in trigonometry). As you can probably imagine, this subject is immensely useful as there are triangles everywhere you go in the world. Trigonometry, for the most part, works only with the lengths, heights, and angles of triangles. This field originated in the 3rd century BC and is now practiced by professionals such as crime scene investigators, engineers, physicists, astronauts, surveyors, and architects.

It may surprise you to learn that trigonometry actually stemmed more from astronomy than any field of mathematics, save geometry. And while trigonometry did not truly develop until around the third century BC, its origins date all the way back to around 2,000 BC.

Now that we have poured over some of the histories of trigonometry briefly, we should now look at some of the subject's everyday uses. Keep in mind that trigonometry's uses are wide and varied, so you cannot expect much depth in these topics for the sake of brevity.

Physics

To start off with, trigonometry can be used in physics to perform a number of tasks. These include finding out the components of vectors that you come across, modeling waves and their mechanics (both physical and electromagnetic waves), modeling oscillations of waves, summing up the strength or weakness of fields (of all sorts), and using dot and cross products. Another principle in physics that trigonometry can be applied to is projectile motion. It is understood that to someone who has no background in physics this all may seem like gibberish. If you want to find out more about these terms there are many great resources on the internet to do so.

Navigation

The next practice that trigonometry can be used for is navigation. Trigonometry was more relevant in this respect to people living in earlier generations without mobile devices but if you are ever stuck without any means of navigation, some basic trigonometry could become useful for you to know.

The first and most important thing that trigonometry is used for in navigation is setting directions such as north, south, east, and west. With this, it can then tell you exactly which direction to take to achieve a straight line if you are using a compass. It can also be used to pinpoint certain locations, and, for mariners, it can be used for finding out a ship's distance from a shore as well.

Criminology

It may come as a surprise to you to learn that trigonometry can also be used in criminology. The most important thing that trigonometry has to offer this field is its ability to calculate a projectile's trajectory through the air. Other important uses of trigonometry in criminology are the estimations it can provide as to might have caused a car crash or how an object fell from somewhere else. Trigonometry can also be used to determine at which angle a bullet has been shot. All of these uses can provide much-needed evidence for police and detectives and can also be used for virtually anyone who wishes to know details on cases.

Marine biology

Another field in which trigonometry is often used that may also come as a surprise to you is marine biology. For example, one of the most common uses of trigonometry within this field is the acquisition of knowledge as to how a level of light affects the ability of algae to photosynthesize. Marine biologists are also known to use mathematical models to measure and analyze marine animals and their behaviors. In addition to these uses, marine biologists also use trigonometry to determine the sizes of larger animals from distances.

Marine engineering

In addition to marine biology, trigonometry is also very useful in marine engineering. In this field, trigonometry is often used to first build and then navigate marine vessels. Specifically, trigonometry can be used to create what is commonly known as the "marine arch", which is a sloping surface that connects higher level areas with lower ones. This forms a triangle, the components of which can be determined by trigonometry alone.

Archaeology
Another field in which trigonometry is commonly used is in archaeology. Archaeologists often use trigonometry to divide their excavation sites into smaller parts. They can also use trigonometry to determine their excavation procedures based on how far they need to dig down to find what they are looking for based on how old what they are looking for is. They also tend to use trigonometry to find out the distance they are from nearby sources of water.

While trigonometry has developed dramatically over the course of its lifetimes, there are a few principles from its inception which are still as valid as ever today. If you were to look back at virtually all of the notable inventions produced since the industrial revolution, you would be hard pressed to find one that does not owe, at least in some measure, its existence to trigonometry. One of the biggest breakthroughs in the history of trigonometry, however, was discovered by Galileo. Galileo discovered that any motion, whatever it may be of, consisted of two components acting on it: the vertical (or gravity) and the horizontal (or the object's projection). He also prognosticated that these two components should always be dealt with independently of one another. These findings caused scientists to gain the ability to measure the velocity of projectiles and the rate at which gravity would act against them, among other abilities.

Surveying

Another practice commonly used in trigonometry is what is known as surveying. One method of doing this is called triangulation which was first suggested by a mathematician by the name of Gemma Frisius. In this method, a person chooses a baseline of a known length, and from this line's endpoints, the angles from it to more remote objects are then measured. This can be done with basic, elementary trigonometry. This process is usually repeated with multiple baselines until the entire area that is being studied is laid out in terms of triangles and their angles. It was a mathematician by the name of Willebrord Snell who first carried out this method on a large scale when he surveyed with 33 triangles an 80 mile stretch of land in Holland.

An even more ambitious survey was then carried out by a French astronaut by the name of Jean Picard, who triangulated the entire nation of France. After that, a yet more ambitious survey was carried out; the entire subcontinent of India was triangulated between the years of 1800 to 1913.

Now that we have gone over some of the everyday uses of trigonometry as well as some of its history, it would be beneficial to go over some of the most basic aspects of making trigonometric calculations.

Trigonometric calculations
The first point that should be touched on is the fact that trigonometry deals primarily with triangles that form right angles, with the sum of all of the internal angles totaling 90 degrees. Having a single right angle in a triangle makes it impossible for all of the sides to reach the same length. The angle opposite

of the right angle is usually labeled as "O". The side opposite of the right angle is what is known as the hypotenuse. The opposite of the angle O, however, is known as the opposite. The other side next to the angle O besides the hypotenuse is known as the adjacent.

Next, we should go over the three main functions in trigonometry when it comes to triangles. Each of these is found by dividing the length of one side of a triangle by another. The first is called sine (or sin for short). To find the sin of a triangle, you must divide the length of the opposite by that of the hypotenuse. The next function is called cosine (or cos for short). To find the cos of a triangle, you must divide the length of the adjacent by that of the hypotenuse. The next and final function is called tangent (or tan for short). To find the tan, you must divide the length of the opposite by that of the adjacent.

SOH CAH TOA

A commonly used abbreviation for finding the value of these functions out is what is known as SOH-CAH-TOA. Or, in equation form, Sin= opposite/ hypotenuse, Cos= adjacent/

hypotenuse, and finally, Tan= opposite/ adjacent.

Let's now look at a real-world example of how all of this could be used. Say you are building a roof and on either side of it the length of the opposite is 20 ft. and the length of the adjacent is 10 ft. Using these figures, we need to look for the hypotenuse's length. TOA will be the equation to use here. To find the tangent here we would take our opposite (10) divided by our adjacent (20). Here are equation would read Tan = 20/ 10, so, tan = 2.

While a calculator is sometimes needed for use on SOH CAH TOA problems with decimals, they can usually be solved by mental calculations with relative ease.

Circles
Next, we should take a look at the uses of trigonometry when it comes to working with circles. The first thing that we should do is divide our circle into four quadrants. At the center of these four quadrants, what is known as a Cartesian coordinate in the center is 0,0. Any point left of the Cartesian coordinate has a negative x value. Any point below the

Cartesian coordinate had a negative y value. So, a point in the upper left coordinate would be (-x, +y), a point in the upper right quadrant would be (+x, +y), a point in the lower left quadrant would be (-x, -y), and finally, a point in the lower right quadrant would be (+x, -y).

Now, if we were to determine a radius within this circle and we were to rotate that radius around the circle we would be left with a series of triangles. This is where our SOH CAH TOA equations come in handy yet again. If the radius was a length of 2 inches and it formed a right triangle with a duplication of the radius and the center of the circle, then the radius would be our adjacent and the duplication of the radius would be our hypotenuse. We would then need to find out the length of the opposite between the adjacent and the hypotenuse. This could be done with the equation CAH. Here we would plug in the adjacent (2 inches), and the hypotenuse (1 inch). So, we would then get 2/1= 2, so the length of the opposite side of the triangle would here be 2 inches.

We should now try out a word problem to get a better grasp on how to use trigonometry in everyday life. Let's say you are out sailing one day but you are not sure where you are going. You initially head out due east, and you do so at a cruising speed around 10 km/ hr. There is, however, also a tide here. It is due north at a speed of 5km/ hr. What direction are you going to end up traveling in under these circumstances?

To answer this question, you must first draw up your triangle. The length of this triangles adjacent would be 10 km, and the length of its opposite would be 5 km. Here we would need to use our tan equation to determine at which angle we are going to sail. This would give us here 5/10= .5.

The inverse tan of .5 is 26.6. This would, however, need to be subtracted from 90 as we are measuring it from 90%. So finally, we would be sailing 63.4% in the northeast direction.

Hopefully, none of these problems included in this chapter required a calculator for you. As you can plainly see by now, trigonometry has many more uses in everyday life than most people expect. If you stick to applying the principles laid out here, you are bound to come across scenarios in your everyday life in which you can apply them. All you have to do is look out for places in which these rules can be applied.

Chapter 3: Adding And Subtracting Fractions

The next point that we are going to touch on here is adding and subtracting fractions. To not be intimidated if you are not a math person here, this chapter is going to be fairly straightforward and simple. It should also be noted that the material in this chapter is some of the most important and useful material that is going to be covered in this book. So, with that being said, it may be helpful to study this chapter in greater depth than you would some of the others.

To start off with, we should define "like fractions". "Like fractions" are fractions that have the same denominators. These fractions can be added and or subtracted with more ease than can other fractions. This is because to add or subtract these you only have to add or subtract their numerators. You would then just carry over the common denominator to your final answer.

If you want to add or subtract fractions with different denominators, you have to start off with finding equivalent fractions with the same denominators. There are two steps in doing this which are listed below:

1.   Find out what the smallest multiple (LCM) is common to both of the numbers.
2.   Next, write out the equivalent fractions with the LCM as the denominators in place of the original fractions.

When you are adding or subtracting fractions, the name for the LCM becomes the lowest common denominator (LCD).

As mentioned before, the addition or subtraction of fractions when their denominators are the same is the easy way. It only becomes difficult to do when denominators are different. Again, when adding or subtracting these fractions you must first find the LCM or, in other words, the LCD. Let's now look at an example of how to find these:

Solving for fractions
Take ¾. To find another way of writing this out we would need to find the lowest number divisible by both of these numbers. In other

words, the smallest number that you can divide both of these numbers by. This number, in this case, would be 12. We would then need to multiply both of these numbers by 12 here. Our equation now would look something like this:

3 x 12 / 4 x 12

This would then equal:

36 / 48

In this example, 36 / 48 would be the equivalent fraction of 3 / 4. This is because both of the terms here have been multiplied by the same number, 12. So, to find equivalent fractions of fractions that do not have common denominators, you must first multiply them by the smallest number which they are both divisible by. Then you can get an equivalent fraction that will have larger terms but will be of the same value.

With that, we will move on to our slightly more complicated second example. This one includes the addition of fractions with different denominators.

¾ + ⅙

The first step we will need to take here is finding the lowest common denominator. This is done most easily by finding all of the multiples of each denominator and seeing which one is the smallest that the two sets have in common. We will now go over this method numerically:

4: 1 x 4= 4, 2 x 4=8, 3 x 4=12, 4 x 4=16
6: 1 x 6=6, 2 x 6=12. 3 x 6=18

As you can see, the first and lowest common multiple of the two is 12, so we would then use 12 as our common denominator.

The second method that we could use in this case is called prime factorization. We would do this by writing out each denominator as a product of its factors. The prime factors of 4 would be 2 and 2. As far as our common denominator is concerned, we need to use the factor which is included in both numbers. We would, therefore, need to use 2 twice and 3 once (3 because of 2 x 3= 6).

So, our prime factorization for 4 would be 2 x 2 and our prime factorization for 6 would be 2 x 3. Meanwhile, our LCD is 12.

Now that we have our lowest common denominator, we need to create equivalent fractions. We would now do this by multiplying each numerator and denominator by their respective factors. Since 3 x 4= 12, we would now multiply 3/ 4 by 3/ 3. Likewise, since 2 x 6 is 12, we would also multiply ⅙ by 2 / 2. These steps would then give us the equivalent fractions 9 / 12 and 2 / 12. At this point, we would need to add up our numerators, 9 + 2. Our answer then would be 11 / 12.

In equation form this would all look something like this:

3 / 4 + 1 / 6 = 3 x 3 / 4 x 3 + 1 x 2 / 6 x 2 = 9 / 12 + 2 / 12 = 11 / 12

Now we should take a look at another example to get a better grasp of what we are doing here.

Using the second method mentioned above, prime factorization, we will now solve for two fractions with mismatching denominators. These are 3 / 10 and 5 / 28.

Next, we would need to find the lowest common denominator.
2 x 5 = 10 and 2 x 2 x 7 = 28, both of these numbers have 2 in common. The number occurs once when multiplying to 10 and twice when multiplying to 28. Now we should take a look at all that we are multiplying and how many times these numbers all occur. Here we are multiplying 2 two times, 5 one time, and 7 one time, so our lowest common denominator would take all of these into account:

2 x 2 x 5 x 7 = 140

Now that we have found our lowest common denominator to be 140 we need to divide this number by both of the original denominators. These equations would look something like this:

140 / 10 = 14, 140 / 28 = 5

Now we would need to create some equivalent fractions from the least common denominator for the two original fractions. Seeing as how the fraction 3 / 10 has a denominator of 10, we would have to multiply it by 14, in this case, to convert it to an equivalent fraction with the lowest common denominator of 140. Our equation for doing so would turn out to look something like this:

3 / 10 = 3 x 14 / 10 x 14 = 42 / 140

Seeing as how 5 / 28 have a denominator of 28, we would have to multiply this fraction by 5 to convert the denominator to those 140 marks. Our equation for doing so would look something like this:

5 / 28 = 5 x 5 / 28 x 5 = 25 / 140

Now that we have our lowest common denominator and we have adjusted our fractions accordingly, our final fractions would wind up looking something like this:

42 / 140
and

Let's now try the same thing with some different fractions. Let's now take 4 / 20 and 5 / 12 and use prime factorization to find their equivalent fractions. We should first start off by finding these two's lowest common denominator. We should now start this off by multiplying up each number's line until we arrive at the lowest common denominator:

20 x 2 = 40, 20 x 3 = 60
12 x 2 = 24, 12 x 3 = 36, 12 x 4 = 48, 12 x 5 = 60

As you can see here, our lowest common denominator, in this case, is 60. Seeing as how 20 x 3 = 60, we would need to multiply 4 x 3 next. This would equal 12. Seeing as how 12 x 5 = 60, we would then need to multiply 5 x 5. This would equal 25. So our final fractions after finding the lowest common denominator and factoring are 12 / 60 and 25 / 60.

Here it would be helpful to go over how to add and subtract mixed numbers. Mixed numbers are values that include both whole numbers and fractions in one. To start off adding or subtracting these, you must first add or subtract the whole numbers and then move on to the additional fractions. Let's start with some easier examples: 4 ⅜ + 2 2/8. This one is easy because the fractions here have common denominators. First we would add up the whole numbers: 4 + 2 = 6, then we would add up the numerators: 3 + 2 = 5. The denominators do not need to be added up in this case because they are the same. Our final answer here would be 6 ⅝.

Another example would be 3 ⅖ + 1 4/5. Again, first, we need to add up our whole numbers and then add up our numerators. The denominators here do not need to be added up because they are the same. So first we would calculate 3 + 1 = 4. And then the numerators, 2 + 4 = 6. 6 / 5 are, however, improper. We would now need to convert this fraction into a mixed number in itself: 6/ 5 = 1 ⅕. So, 4 6/5 = 4 + 1 ⅕ = 5 ⅕. Our final answer here would be 5 ⅕.

Our next example is more complicated and it features denominators that are different from one another. The basic equation is as follows: 6 ¾ + 3 ⅝

Because these two denominators are different we need first to find the lowest common denominator here. Between 4 and 8 the lowest common denominator would be 8, so we would not need to multiply the numerator of the first fraction but we would need to multiply the first numerator, 3, by 2 because we had to multiply 4 by 2 to get to 8. So we would then come up with 6 6/8 + 3 ⅝. This would give us 9 11/8. 11/8, however, is an irregular fraction which we would need to convert to 1 ⅜. So now we are left with 9 + 1 ⅜ = 10 ⅜

Let's now look at another example of adding mixed numbers in which the denominators are different from one another, 8 2/4 + 6 ⅜.

The lowest common denominator here is 8. Since we would need to multiply 4 by 2 to arrive at 8, we would then need to multiply the first denominator, 2, by 2 as well. This would give us 4. Our new equation would then look like this: 8 4/8 + 6 ⅜. First, we would need to add up the whole numbers which would give us 14. Then, we would need to add up our numerators, which would give us 7. Our final answer here would be 14 ⅞.

As you can probably tell by now, adding and subtracting fractions and mixed numbers is not only easier than it may seem at first but it is also incredibly useful in everyday life. You deal with fractions all the time in life whether you are cooking from recipes, balancing your checkbook, or finding out novel statistics on the topics that interest you. You might as well learn how to deal with problems concerning fractions better when considering just how much of your time is spent dealing with these.

Mixed numbers are not all that complicated to deal with either and they prove to be very useful when dealing with difficult fractions that should not stand out alone by themselves. Upon investigating these you should find that mixed numbers are also very common and very easy to solve for. All of the principles laid out within this chapter should be feasible to follow mentally, tough if you feel that going over them while taking notes will in any way help you, you should try to do that.

Chapter 4: Mean, Median, And Standard

Deviation In Everyday Life

Averages are extremely useful in that they tell us what the normative figures in a sequence are. There are many different types of averages, all of which give us different and unique perspectives on what is going on in the center of our distributions. Of these types of central measurement, the main ones are median, mean, average, mode, and range. This chapter will be dedicated to the study of these measurements.

Mode, median, and mean are all different types of averages. Averages are very common in any type of statistics but these three are usually the most common types that you will come into contact with. Of these three, the mean is probably the one which is most commonly calculated to by most people. To find the mean in a set of data, you must first add up all the numbers presented to you and then divide that sum by the amount of numbers in the set of data. The median is the middle of all of the numbers. This is usually much easier to find than the mean in any given data set. Just keep in mind that your numbers have to be arranged from smallest to largest when accounting for the median, so you may need to do some rearranging of figures before you are able to arrive at your number. The mode is the only one of these three that has a possibility of not occurring at all throughout a data set. The mode is simply the number that occurs most often throughout the set of data. If no numbers are repeated, or none are repeated any more than any others, then there will be no mode for the set of data. In addition, a range is a difference between the largest and smallest numbers within a set of data.

We should now start on some examples of value sets and how to solve for their averages. Let us now find the range, mode, median, and mean for this value set:
Findings averages in data sets

13, 18, 13, 14, 13, 16, 14, 21, 13

We should start off by calculating the mean, as it is the most common average. We would do this here by adding up all of the terms and dividing by the number of all of the terms. This would look something like this in equation form:

$(13 + 18 + 13 + 14 + 13 + 16 + 14 + 21 + 13) \div 9 = 15$

So, our mean has turned out to be 15 in this case. You may have noticed that this number was not included in the original set of terms. This is common when finding the mean. They sometimes do not occur within the original data set.

Next, we need to find the median. The median is the middle value within the number set, so in order to find this, we first need to reorder our dataset from the lowest to the highest numbers. This would look something like this:

13, 13, 13, 13, 14, 14, 16, 18, 21

As you can see here, there are nine numbers included within this set. This means that our median would be found in the 5th number listed here because that is the halfway point between 9 and 1. In this case, the 5th number listed is 14, so our median here would be 14.

Next up we would need to find the mode. The mode represents the number which is most commonly repeated, so within this set-out mode would be 13.

And at long last, we come to our range. Again, this is calculated by subtracting the smallest number from the largest number. Within this set, 21 is our largest number while 13 is our smallest number. So our equation would read: 21 - 13 = 8. Our range, in this case, would be 8.

So, to recap, our mean here is 15, our median 14, our mode 13, and our range here is 8.

Let's now look at another example to get a better grasp on how to find these averages.

Let's take the number set: 1, 2, 4, and 7 for example.

First off, we would need to calculate the mean by adding up all the numbers within the set and then dividing then all by the number of terms, 4 in this case. This would look something like this in equation form: (1 + 2 + 4 + 7) ÷ 4 = 14 ÷ 4 = 3.5

As you can see, our mean here is 3.5 which is an irregular number. This is common when finding the mean.

Next, we need to find out what the median is. The median is the middle number within a data set. Here we do not need to reorder out numbers because they are already in numerical order. There is, however, no number in the middle here because this set has an even amount of terms. When this occurs we need to find out the halfway point between the two middle terms which are in this case 2 and 4. Our median would, therefore, be 3 in this case. 3 are, as you can see, not on this list at all, which will happen sometimes when you are finding the median.

Next, we need to find the mode. The mode is repeated oftentimes, however, none of the numbers listed in the set are ever repeated, so we would, in this case, have no mode.

And finally, we need to find the range of this set. This is done by subtracting the smallest number from the largest. Within this set, our smallest number is 1 while our largest number is 7. 7 - 1 = 6, so our range, in this case, would come out to equal 6.

To recap, our mean for this data set is 3.5, our median here is 3, we would have no mode under these circumstances, and finally, our range here is 6.

This next example that we will now go over has many different characteristics than the ones that we have gone over previously. Let's now start with our standard set of numbers: 8, 9, 10, 10, 10, 11, 11, 11, 12, and 13.

To start off with, we first need to find the mean. This is, again, only done by adding up all of our terms within a set and then dividing the sum by the number of terms. In equation form this would all look something like this: (8 + 9 + 10 + 10 + 10 + 11 + 11 + 11 + 12 + 13) ÷ 10 = 105 ÷ 10 = 10.5

So, as you can see, our mean here is 10.5, which is an irregular number, but this is commonly the case with the mean.

Next, we need to find the median. The median, again, is the middle number within any given set of data. Here we have 10 terms in our set which is an even number. This means that there is no number in the middle of all of these. We would, therefore, need to find the value between the two innermost terms, 10 and 11. This would give us a median of 10.5 which is like our mean, an irregular number but this is common in finding the median.

Next, we would need to find the mode of this set. The mode is, again, the number which occurs most commonly within a set of data. Here we have two modes. These are 10 and 11 because they both occur three times each which is more common than any of the other terms listed here. Having two or more modes is another commonality in finding averages.

And finally, we come to find the range. We would need to find the range by subtracting the lowest term in the set from the largest term in the set. In equation form, this would look something like this: 13 - 8 = 5

So, as you can see, our range here would be 5.

To recap, our mean for this set turned out to be 10.5, our median is also shown here to be 10.5, we would have two modes in this example which would be 10 and 11, and our range is 5.

The mean and the median, in this case, turned out to be of the same values. This can occur often when finding out averages despite which averages the similarities happen between. When you find that two or more averages are of the same value within a set, do not take it as evidence that you did your calculations wrong as that will happen from time to time naturally.

Next, we should look at a word problem concerning averages. This will give us a better grasp of how to apply these concepts to real, everyday life.

The following are the test scores that a student got on the most recent tests: 87, 95, 76, and 88. He is hoping to get at least an 85 or better in the class. Considering these previous scores, what grade will he need to get on his final test at a minimum in order to meet his goal?

What we are trying to solve for here is the minimum grade that he needs to get in order to achieve this. First, we would need to find the average of his test scores by mean, or adding up all of his test scores and dividing the sum by the overall number of tests. Since the score for the last test is what we are trying to solve for here, this will be represented by x within the equation: $(87 + 95 + 76 + 88 + x) \div 5 = 85$

Once we have multiplied by 5, the equation then turns into this: $87 + 95 + 76 + 88 + x = 425$ which would in turn become this: $346 + x = 425$.

X would then equal 79, so he would need to get at least a 79% on his final test in order to achieve his goal of an average of 85% in the class overall.

Let's now take a look at another word problem.

A woman has four relatives. Their ages are 45, 36, 60, 18, and 20. What are the mean, median, mode, and range of this set of ages?

First, we need to calculate for the mean by adding up all of the numbers and then dividing the sum by the number of terms listed. In equation form this would look something like this: $(45 + 36 + 60 + 18 + 20) / 5 = 179 / 5 = 35.8$

As you can see, our mean here is irregular, which again is fine and normal.

Next, we would need to find our median. This is the number in the middle of the set, which in this case would be 36 numerically, so our median here is 36.

We would have no mode within this set because the mode is the most frequently occurring number within a set. None of the numbers listed above recur at all.

And finally, we need to find the range of this set. This is done by subtracting the smallest number from the largest number. As an equation, this would look something like this: 60 - 18 = 42. Our range, in this case, would, therefore, be 42.

As you can probably tell by now, finding out the values of the various types of averages within sets of data is a fairly straightforward process which does not provide very much difficulty once you initially learn how to perform all of these operations. Once you master these skills in making these calculations, you will invariably find that finding averages comes in handy all the time throughout your everyday life. There are averages within every set of numbers that you come across, so now that you have the tools to find these averages, you should use them to your full advantage when you find opportunities to do so.

Chapter 5: Working With Conversion Factors

The next topic that we should go over is conversion factors. These are incredibly useful for anyone because we are in more or less constant contact with them every day. They come in all forms, some of which are easy to adjust to; others are more complicated and or vague in their conversions. Whatever way in which you find them, they come in handy when making everyday calculations nonetheless. We will now start out by going over some more esoteric conversion factors which most people do not come into contact with on a regular basis.

Conversion factors
To convert acres to hectares you need to multiply by .4047. To convert acres to square feet, you need to multiply by 43,560. To convert acres to square miles, you need to multiply by .001562.

To convert atmospheres to centimeters of mercury, you need to multiply by 76. To convert Btu/hour to horsepower, you need to multiply by .0003930. To convert Btu to kilowatt-hour, you need to multiply by .0002931. To convert Btu/hour to watts, you need to multiply by .2931.

To convert bushels to cubic inches, you need to multiply by 2150. To convert 4 bushels (U.S.) to hectoliters, you need to multiply by .3524. To convert centimeters to inches, you need to multiply by .3937. To convert centimeters to feet, you need to multiply by .03281.

To convert cubic feet to cubic meters, you need to multiply by .0283. To convert cubic meters to cubic feet, you need to multiply by 35.3145. To convert cubic meters to cubic yards, you need to multiply by 1.3079. To convert cubic yards to cubic meters, you need to multiply by .7646.

To convert degrees to radians, you need to multiply by .01745. To convert dynes to grams, you need to multiply by .00102. To convert fathoms to feet, you need to multiply by 6. To convert ft. to meters, you need to multiply by .3048. To convert feet to miles (nautical), you need to multiply by .0001645. To convert feet to miles (statute), you need to multiply by .0001894. To convert feet/second to miles/hour, you need to multiply by .6818. To convert furlongs to feet, you need to multiply by 660.0. To convert furlongs to miles, you need to multiply by .125.

To convert gallons (U.S.) to liters, you need to multiply by 3.7853. To convert grains to grams, you need to multiply by .0648. To convert grams to grains, you need to multiply by 15.4324. To convert grams to ounces (avdp), you need to multiply by .0353. To convert grams to pounds, you need to multiply by .002205.

To convert hectares to acres, you need to multiply by 2.4710. To convert hectoliters to bushels (U.S.), you need to multiply by 2.8378. To convert horsepower to watts, you need to multiply by 745.7. To convert horsepower to Btu/hour, you need to multiply by 2,547.

To convert hours to days, you need to multiply by .04167. To convert inches to millimeters, you need to multiply by 25.4000. To convert inches to centimeters, you need to multiply by 2.5400. To convert kilograms to pounds (avdp or troy), you need to multiply by 2.2046. To convert km. to mi., you need to multiply by .6214.

To convert kilowatt-hour to Btu, you need to multiply by 3412. To convert knots to nautical miles/hour, you need to multiply by 1. To convert knots to statute miles/hour, you need to multiply by 1.151.

To convert liters to gallons (U.S.), you need to multiply by .2642. To convert liters to pecks, you need to multiply by .1135. To convert liters to pints (dry), you need to multiply by 1.8162. To convert liters to pints (liquid), you need to multiply by 2.1134. To convert liters to quarts (dry), you need to multiply by .9081. To convert liters to quarts (liquid), you need to multiply by 1.0567.

To convert meters to feet, you need to multiply by 3.2808. To convert meters to miles, you need to multiply by .0006214. To convert meters to yards, you need to multiply by 1.0936. To convert metric tons to tons (long), you need to multiply by .9842. To convert metric tons to tons (short), you need to multiply by 1.1023. To convert miles to kilometers, you need to multiply by 1.6093.

To convert miles to feet, you need to multiply by 5280.

To convert miles (nautical) to miles (statute), you need to multiply by 1.1516. To convert miles (statute) to miles (nautical), you need to multiply by .8684. To convert miles/hour to feet/minute, you need to multiply by 88.

To convert millimeters to inches, you need to multiply by .0394. To convert ounces (avdp) to grams, you need to multiply by 28.3495. To convert ounces to pounds, you need to multiply by .0625. To convert ounces (troy) to ounces (avdp), you need to multiply by 1.09714. To convert pecks to liters, you need to multiply by 8.8096. To convert pints (dry) to liters, you need to multiply by .5506. To convert pints (liquid) to liters, you need to multiply by .4732.

To convert pounds (ap or troy) to kilograms, you need to multiply by .3732. To convert pounds (avdp) to kilograms, you need to multiply by .4536. To convert pounds to ounces, you need to multiply by 16. To convert quarts (dry) to liters, you need to multiply by 1.1012. To convert quarts (liquid) to liters, you need to multiply by .9463.

To convert radians to degrees, you need to multiply by 57.30. To convert rods to meters, you need to multiply by 5.029. To convert rods to feet, you need to multiply by 16.5. To convert square feet to square meters, you need to multiply by .0929. To convert square kilometers to square miles, you need to multiply by .3861. To convert square meters to square feet, you need to multiply by 10.7639. To convert square meters to square yards, you need to multiply by 1.1960. To convert square miles to square kilometers, you need to multiply by 2.5900. To convert square yards to square meters, you need to multiply by .8361.

To convert tons (long) to metric tons, you need to multiply by 1.016. To convert tons (short) to metric tons, you need to multiply by .9072. To convert tons (long) to pounds, you need to multiply by 2240. To convert tons (short) to pounds, you need to multiply by 2000.

To convert watts to Btu/hour, you need to multiply by 3.4121. To convert watts to horsepower, you need to multiply by .001341. To convert yards to meters, you need to multiply by .9144. To convert yards to miles, you need to multiply by .0005682.

As you can probably already tell, there are no great methods of memorizing these. You can only master them by rote memorization which is not very efficient but will still give you the greatest results possible in this case.

When thinking about conversion factors it is useful to see them as a means of changing the units of a measured quantity without changing its value. There is one method of calculating the value of a conversion factor. This is called the unity bracket method. The unity bracket method involves setting the numerator and denominator of a fraction at the same value in order to determine the applicable unit conversion factor.

Conversion factors are, as you can probably already tell, very easy to deal with, they just take time to memorize. Once these are memorized, however, they do very much come in handy in everyday life and are usually easily committed to long-term memory. The only hard part of dealing with these factors is the initial memorization of certain multiplications that is necessary. Keep in mind also that you do not need to memorize all of the conversion factors mentioned above. You would actually be wiser to only memorize the ones that you expect are going to be immediately useful to you in your everyday life.

Using the PIN Technique

Next, it would be helpful for us to go over the plugging in numbers or PIN technique. This technique is especially helpful for students because it offers a fast and easy way to make calculations without any calculator. On the ACT and the SAT tests, as well as any other standardized tests which include math, the only thing that matters to your graders is that you arrived at the correct answer. It is not relevant by what means you came to the answer. This gives you much-needed flexibility as you do not necessarily have to stick with the methods taught to you by your teachers in order to pass these tests. Below we are going to go over why and how to use the PIN method when making your calculations.

First, we should start out with why this method is useful. There are bound to be a lot of problems that you simply do not know how to approach, have too many variables for you to know their right answer, or seem like they would take too long for you to solve. These circumstances are where the PIN method can prove to be rather useful.

Let's first take solving problems with lots of variables for example. This can seem intimidating and overly time consuming at first but if you replace all of the variables within these problems with certain numbers it can make the whole process a lot more bearable and a lot less time-consuming. It will also make these problems look a lot more accessible and a lot less confusing. Let's now look at an example of one of these problems in which we would apply the PIN method:

The PIN method
For all numbers that represent a and b, let a b be defined as a b = ab + a + b. For all numbers

that represent x, y, and z, which of the following is true?

1. x y = y x

2. (x - 1) (x + 1) = (x x) - 1

3. x (y + z) = (x y) + (x z)

Answers: a) 1 only, b) 2 only, c) 3 only, d) 1 and 2 only, e) 1, 2, and 3.

This answer will be brought up below.

It can be easy to forget that you have this method at your disposal when you are taking a test. Keep in mind though that this method will help you to simplify the problems that you come across greatly. Whenever you see a problem on a test that includes more variables than you are comfortable with, you should keep this method in mind for such occasions.

Now that we have covered why the PIN method is helpful, we should delve into how to go about using it. The basic idea of the PIN method, as mentioned before, is that you replace the variables that you are faced with other numbers. The technique can work for any problem in which you are confronted with variables whether it is geometry or algebra that you are doing.

A general rule of thumb for using the PIN method is that you can use it whenever there are variables in the equation. The more variables that an equation has, the more useful the PIN method becomes.

Because these problems involving variables determine relationships between numbers, you can tell within these which relationships are constant and which are not. In other words, you can see which relationships hold true regardless of the numbers being used. As long as your own numbers follow the rules set forth within the equation, you can use your own numbers in replacement of the variables originally set forth and arrive at the correct answer.

Once you have picked your desired number to represent the variable, you can then solve the original equation with that number. After that, you can then look for the original variable in your answer options and replace it with the new number that you have chosen. By doing this you can more easily test your options as far as answers go and see which of the answers best matches the results you came up with for the problem. If this does not make total sense yet, do not worry. We are now going to go over the previous problem in more detail to get a better grasp of what we are doing here.

For all numbers that represent a and b, let a b be defined as a b = ab + a + b. For all numbers that represent x, y, and z, which of the following is true?

1. x y = y x

2. (x - 1) (x + 1) = (x x) - 1

3. x (y + z) = (x y) + (x z)

Answers: a) 1 only, b) 2 only, c) 3 only, d) 1 and 2 only, e) 1, 2, and 3.

We are told that the relationship mentioned above applies equally to all of the numbers x, y, and z. We are therefore able to replace any of these letters x, y, and z with any given numbers here because any and all numbers would work in this case.

We should now give each of these variables their own numbers. Seeing as how they are all different, they should each get their own individual numbers assigned to them. This would look like this:

x = 2, y = 3, z = 4

Let's now solve this problem using the new numbers that we have assigned to these variables and see if everything works out the same for us using this method.

The first part is:

x (+) y = y (+) x

We should now take half of this equation and replace the variables with our new numbers:
2 (+) 3

This would in turn become:

(2) (3) + 2 + 3 = 11

Now that we see that the left half of our equation equals 11 we should solve for our right half:

$$y (+) x = 3 (+) 2 = (3) (2) + 3 + 2 = 11$$

So, as you can now see, both of our halves equal one another in this case. Now we will refer back to our answers:

1. $x y = y x$

2. $(x - 1) (x + 1) = (x x) - 1$

3. $x (y + z) = (x y) + (x z)$

Answers: a) 1 only, b) 2 only, c) 3 only, d) 1 and 2 only, e) 1, 2, and 3.

Now that we know that both sides equal 11, we can deduce from that that the first statement is true. This, in turn, means that we can disregard b and c as possible answers because they exclude the first statement listed here.

We should now try the PIN method on statement to see if it is also true. Let's first take a look at the original equation:

$$(x - 1) (x + 1) = (x x) - 1$$

Again, we should start by applying our numbers to the left side of the equation first. This would look something like this:

$$(x - 1) (+) (x + 1) = (2 - 1) (+) (2 + 1) = 1 (+) 3 = (1) (3) + 1 + 3 = 7$$

So now that we know that the left side of out equation equals 7, we should move on to the right side of the equation. The original equation on the right side looks like this:

$$((x (+) x)) - 1$$

Now we would have to replace the xs here with our replacement number, 2.

$$((2 (+) 2)) - 1 = ((2) (2) + 2 + 2) - 1 = 7$$

So here our left side equals 7, as does our right side as well. This means that statement number 2 is also correct. We can then eliminate answer a in addition to answers b and c.

Our final step would be to replace the variables in the last statement with the numbers that we have decided on. The original equation here looks something like this:

$$x (y + z) = (x y) + (x z)$$

First, we need to plug in our substitute numbers into the left side of the equation. After doing so the left side equation turns into this:

$$2 (+) (3 + 4)$$

Which would then become:

$$2 (+) 7 = (2) (7) + 2 + 7 = 23$$

So, as you can see, the left side of this equation equals 23. Next, we need to solve for the right side in order to see if the result matches the answer for the left side. The original equation for the right side looks something like this: (x y) + (x z).

After substituting the variables with our new numbers we would come up with this:

$$((2 (+) 3)) + ((2) (+) 4)) = ((2) (3) + 2 + 3) + ((2) (4) + 2 + 4) = (7) + (14) = 25$$

So our answer for the right side of the equation here is 25. Our answer for the left side of this equation was, again, 23, which means that the two sides are not equal and therefore statement number 3 is not correct in this case. This would leave us with answer d as the only correct answer. Statements 1 and 2, in this case, are correct statements, while statement 3 is incorrect.

We were able to choose all of our own numbers in this example, but do keep in mind that this is not always the case when using the PIN technique. Always be on the lookout for when you can choose your own numbers for most, if not all, of the variables in a given equation or you can choose your own number for one of the variables and solve for the rest. This will give you a very helpful advantage in solving problems with more efficiency. We were able to choose our own numbers for all of the variables in the problem mentioned above only because we were told that the variables applied to all numbers in that case. Any number that we could have chosen would have followed the rules laid out for us.

You should only assume that you are able to plug in your own numbers in place of variables when the problem tells you specifically that the variables apply to any and all numbers. Usually, when you do not see this directive written out explicitly, you are allowed to choose your own number for one variable while still solving for all or the others. This will ensure that all of the variables are following their own rules and are keeping the relationships between one another intact.

We should now solve for an example of a problem in which we do not have the option of determining all of our own variables.

$x = 3v, v = 4t, x = pt$

Given the set of equations listed above, if $x = 0$, what is p's value?

We are not told here that this problem applies to any and all numbers which are why we can only make our substitution for one variable and leave the rest alone. To solve here we are going to replace v with our own number. This is because v shows up in the middle equation, which shares the variables of the other two equations.

The first thing that we should note now is the fact that v = 4t. With this in mind, it becomes clear that we should choose a number divisible by 4 to take the place of v., In this case, let's now say that v = 8. So now if we replace every v in every equation our first equation will then look like this:

$x = 3 (8) = x = 24$

So, as you can now see, when v = 8, x = 24 in turn. Now we need to move on to our next equation. The original version looks like this: v = 4t

When substituting for v we would then get

$8 = 4t = t = 2$

So, when x = 24 and v = 8, t = 2 in turn.

Our final step would be to now take a look at the last equation using our newfound values for the variables. The original equation looks like this:

x = pt

When the variables are substituted this equation turns into this:

24 = p(2) = p = 12

So, as you can see, p would equal 12 in this case. You could now make the assertion that p may not equal 12 in this case if v had not been originally determined to equal 8. We could then test this assertion by given v a different value. Let's now say that v = 20.

Our first equation under these circumstances would turn into this:

x = 3 (20) = x = 60

Our second equation would then turn into this:

$20 = 4t = t = 5$

And our final equation would then turn into this:

$60 = p(5) = p = 12$

So, as you can see, p again equals 12 here. This is because we have kept our variables intact while solving both times. It is not relevant what any of the other variables are determined to be so long as we keep these variables intact. Our final answer, in this case, would be 12, or:

$p = 12$

So those are the bigger points on using the PIN strategy. Now that you have some of the skills concerning the use of this strategy, it should be easy to apply these when you are taking any tests with problems that may require this strategy in the future. The PIN strategy is also helpful in everyday life as well. It will come in handy whenever you have problems with too many variables for you to be comfortable with. Keeping this strategy in mind in your future as well as its application can give you a great advantage in math and general problem-solving.

Chapter 6: Techniques For Sat, Gmat, And

Gre Students

This next chapter will be devoted exclusively for use by SAT, GMAT, and GRE students. We will focus mainly on SAT practice questions here because these are similar enough to the questions of the other tests to still be useful for students preparing for the GMAT or the GRE. Feel free to skip over this chapter if you are not studying for any of these tests and feel that this will not be a great use of your time.

To that end, our first practice question is an algebraic word problem.

SAT problems
Within a classroom at central high school, the mean number of students (or y) can be determined by the equation $y = 0.8636x + 27.227$. X, in this case, represents the number of years it has been from 2004, it is less than or equal to 10. Of the following statements, which is the best interpretation of the

significance of the number 0.8636 within the context of this classroom?

a.   The mean number of students in the classroom in the year 2004.
b.   The mean number of students in the classroom in the year 2014.
c.   The yearly decrease in the mean number of students in the classroom.
d.   The yearly increase in the mean number of students in the classroom.

To answer this question, we would need to determine the slope of the equation and its relationships to the real world situation it models. You should also keep in mind that we are only solving for the independent variable, y, here instead of the dependent variable, x.

Choice d here would be the correct one. Let's now determine why this is. When this equation is written in y= mx+ b, or slope form, the coefficient of x (or 0.8636 in this case) would be what is known as the slope. The slope of this equation would then give you the amount by which the mean number of students in the classroom changes on average each year. The slope here would be a positive one which means that there is an increase in the mean number of students each year.

Let's now move on to our next example:

If 2/a − 1 = 4/y, and y ≠ 0 where a ≠ 1, what is y in terms of a?

a)   y = 2a − 2
b)   y = 2a − 4
c)   y = 2a − 1 / 2
d)   y = 1 / 2 a + 1

Choice a is the correct one in this case. Now we will go over why it is.

First, you need to cross multiply the denominators by their opposite numerators which would be 2 x y = 2y, and 4 x (a - 1) = 4a - 4.

So, as you can see, this would give you the equation

2y = 4a - 4

To solve this you would then divide both sides by 2 in order to isolate the variable y.

So, $y = 2a - 2$ in this case. Therefore, the correct answer would be a here.

Reliable SAT, GMAT or GRE study would have to go very far beyond the problems mentioned here in order to be effective. These are very long and expansive tests that need to be taken seriously with long hours of devoted and focused study beforehand by those who take them. There are innumerable resources on preparing for these tests, so if you have one of these coming up it would be very beneficial for you to look into more on this subject. It should also be noted that we have only covered some of the algebra questions featured on the test here. The actual test itself goes over many other types of math than just algebra.

These tests also involve sections in which students are tested upon other subjects, so if you are about to take one of these tests, make sure that you are studying an array of subject matter in order to be prepared for all dimensions of the test at hand.

Math Strategies for Anyone

This book is, however, not just meant for use by students preparing for exams. This book is written for anyone who desires better mental math skills. Now we should go over some of the most basic aspects of mental math that you should keep at your disposal at all times in order to be better able at making calculations on the fly.

One aspect of mathematics that any adult can appreciate and apply to his or her everyday life is adjusted gross income. This is defined as the sum of the income that a person acquires within a year. Let's now look at an example of how to solve for adjusted gross income.

Adjusted gross income
In 2009, a man named Peter made a total of $15,000 working at his main job. In addition, he also made 200 extra dollars working another job for 20 weeks. Using these figures, we now need to find Peter's adjusted gross income.

Earnings = weekly earning × number of weeks worked

Earnings = 200 × 20 = 4,000

Adjusted gross income = wages + interest income

Adjusted gross income = 15,000 + 4,000 + 50
Peter's adjusted gross income would, therefore, be = 19,050 USD.

In 2009, a woman named Martha earned a total of $80,000 at her primary job. In addition, she also earned $50,000, $20,000, and $100,000 from her side jobs. Using these figures, we now need to determine Martha's adjusted gross income.

What is Martha's adjusted gross income?

Adjusted gross income = wages + interest income

Adjusted gross income = 80,000 + 50,000 + 100,000 + 20,000

Martha's adjusted gross income would, therefore, be = 250,000 USD.

Next, we should take a look at how to calculate for price discounts. Price discounts are when the price of an item is marked down, usually by a certain percentage. Here we will go over how to determine the new price of an item based off of the list price and the discount rate of the item.

Price discounts
Here is our first example: we need to find the new price of an item with a list price of 100 dollars and a discount rate of 25%.

Discount = 100 × 25% = 100 × 0.25 = 25

Sale price = List Price - Discount Price = 100 - 25 = 75 dollars

And on to our second example: here we need to find the sale price of an item that has a list price is 24 dollars and a discount rate of 50%.

Discount = 24 × 50% = 24 × 0.50 = 12

Sale Price = List Price - Discount = 24 – 12 = 12 dollars

Now we should go over some math for investment strategies. Here we will delve into calculating for simple interest. In its simplest form, the equation for simple interest is as follows: Interest = Principal × Rate of Interest × Time

Simple interest
We should now look at an example of how to apply this equation. Here we need to compute the interest if the principal is 2,000,000 dollars at a rate of interest of 4% for a year.

We may need to use a calculator here:

Interest = 2,000,000 × 4% × 1

Interest = 2,000,000 × 0.04 × 1

Interest = 80,000 × 1 = 80,000

And here is another example of calculating simple interest in real life. Here we need to compute the interest if the principal is 100 dollars at a rate of interest of 2% for 10 years.

Using a calculator:

Interest = 100 × 2% × 10

Interest = 100 × 0.02 × 10

Interest = 2 × 10 = 20

Next, we will go over mortgage loans. This advice will be particularly useful for anyone considering buying a home.

Mortgage loans
In this example, you are buying a house for $250,000. Towards this, you initially make a down payment of 15% of the purchase price and then you agree to a 30-year mortgage to cover the balance.

With all of this in mind, what is your down payment? And what is your mortgage?

Down payment = Purchase Price × Percent Down

Down payment = 250,000 × 0.15 = 37,500

Amount of Mortgage = Purchase Price – Down Payment

Amount of Mortgage = 250,000 − 37,500 = 212,500

If let's say, your monthly payment is 1,200 dollars, what is the total interest charged over the life of the loan?

Total Monthly Payment = Monthly payment × 12 Months per year × Number of years

Total Monthly Payment = 1,200 × 12 × 30 = 432,000

Total Interest Paid = Total Monthly Payment − Amount of Mortgage

Total Interest Paid = 432,000 − 212,500 = 219,500

This concludes our consumer math section of the book. The section should be immediately useful for most people than are any others. If you apply the equations mentioned here the next time that you find yourself dealing with mortgage loans, simple interest, discount prices, or adjusted gross income, you will be better able to make decisions regarding these matters.

Chapter 7: Addition Techniques

Make the distinction between a number and a digit

Later in this book, we are going to talk about digits and numbers. That is why, in order for you to follow me, it is important to clarify the difference between these two words:
A digit goes from 0 to 9: so there are 10 digits, which are 0, 1, 2, 3, 4, 5, 6, 7, 8 and 9.

A number consists of a group of end-to-end digits: for example, 1256 is a 4-digit number, composed of the digits 1, 2, 5 and 6. Same way 658954 is a 6-digit number, composed of the digits 6, 5, 8, 9, 5 and 4.

If that is clear, we can move forward to the next stage.

The key: add from left to right

Most of the calculation techniques taught at school are adapted to written problem solving. In this way, it is common to do operations from right to left. For all that, daily life situations are more suitable for mental arithmetic.

Yet, one of the first pieces of advice we can give to anyone wishing to develop his mental arithmetic capacities is getting into the habit of doing calculations from left to right.

In writing, it is possible to add 4500 + 67 from right to left; however, when the calculation is done mentally, it is more natural and faster doing it from left to right.

This advice applies whatever the operation type you wish to do: addition, subtraction, multiplication, and division.

In mental arithmetic, always solve operations from left to right.

Add faster by the regrouping method

You would more easily solve additions if you could identify digits that, added together, give 10. Digits to associate are:

1	9	1 + 9 = 10
2	8	2 + 8 =10
3	7	3 + 7 = 10
4	6	4 + 6 = 10
5	5	5 + 5 = 10

Figure 1 : Regrouping table

So, when solving an addition, begin by mentally grouping numbers that end in complementary digits.

Illustration:

Your shopping list includes 6 articles, as pointed below:

Toothpaste 7 €
Shower gel 6 €
Meats 31 €
Vegetables 24 €
Drinks 23 €
Frozen food 19 €

Solving the addition of this cashier ticket, group articles according to its price's last digit, as pointed out in the regrouping table (figure 1), which result in:

Toothpaste   7 €
Drinks 23 €
Shower gel   6 €
Vegetables   24 €
Meats 31 €
Frozen food   19 €

In this way, it is easier to solve additions 2 by 2 as:
toothpaste + drinks = 7 + 23 = 30 €
shower gel + vegetables = 6 + 24 = 30 €
meats + frozen food = 31 + 19 = 50 €

That is to say a total of 110 €.
Exercises:
Do the regroups, allowing you to facilitate the calculation of:

a/ 46 + 25 + 53 + 4 + 37 + 15
b/ 12 + 21 + 14 + 39 + 16 + 28
c/ 105 + 33 + 60 + 10 + 25 + 47

Answers:

a/ (46 + 4) + (25 + 15) + (53 + 37) = 50 + 40 + 90 = 180

b/ (12 + 28) + (21 + 39) + (14 + 16) = 40 + 60 + 30 = 130

c/ (105 + 25) + (33 + 47) + (60 + 10) = 130 + 80 + 70 = 280

Decompose an addition with close numbers

The regrouping method shows that it is easier to solve additions as soon as we see numbers containing 0 in the calculation.

In this way, when some addition numbers are close to whole numbers such as 100, 200, 300... 1000 etc..., it is very useful to handle these numbers in order to considerably simplify the calculation solving.

The game objective here is to modify the addition in order to write it in a more legible and more explicable way for our brain.

Illustration:

Suppose that we have this addition to solve: 356 + 696.

Mentally, you identify that 696 is close to 700: 696 = 700 - 4

The proposed addition can then be written in a simpler way, that is: 356 + 700 - 4

Calculating from left to right: 1056 − 4 = 1052

Use the same principle to calculate: 204 + 387 + 615

Mentally, you identify that:
204 is close to 200 (204 = 200 + 4)
387 is close to 400 (387 = 400 − 13)
615 is close to 600 (615 = 600 + 15)

We can then rewrite the addition this way:
200 + 4 + 400 − 13 + 600 + 15

Using the regroup method: 200 + 400 + 600 + 4 − 13 + 15

That is to say 1200 + 6 = 1206

Exercises:

Solve these next additions by the decomposition method:

a/ 196 + 742
b/ 203 + 536 + 488

Answers:
a/ 200 − 4 + 742 = 942 − 4 = 938
b/ 200 + 3 + 500 + 36 + 500 − 12 = 1200 + 27 = 1227

Use the decomposition method by close numbers to simplify additions making appear 0.

Cut out numbers to solve an addition: guaranteed performance!

A little known but very efficient method to simplify a difficult addition into 2 or 3 simpler additions consists of cut out numbers.

When mentally calculating, cutting out numbers can considerably simplify the calculation's difficulty.

Illustration:

Suppose you have this addition to solve: 326521 + 432478.
Mentally, you undertake the next cutting out:
32 / 65 / 21
43 / 24 / 78

And you add numbers cut out this way:
32 + 43 = 75
65 + 24 = 89
21 + 78 = 99

Which results in putting the results end-to-end: 758999.

Sometimes, calculation is a little more complicated because there is a remainder as when you have to calculate 541268 + 39323

Mentally, you undertake the next cutting out:
54 / 12 / 68
03 / 93 / 23 (we add a 0 at the beginning in order to make sure the 2 numbers to be added each have 6 digits)

And you add numbers cut out this way:

54 + 03 = 57

12 + 93 = 105 (you should keep only 2 digits, the 1 will add to the previous 57, what results in 58)

68 + 23 = 91

What results in 580591

Exercises:

Solve next additions by the cutting out method:

a/ 541247 + 251478
b/ 32569 + 4781
c/ 365214 + 325874

Answers:
a/ 54/12/47 + 25/14/78 = 54 + 25 / 12 + 14 / 47 + 78 = 79 / 26 / 125 = 79 / 27 / 25 = 792725
b/ 3/25/69 + 47/81 = 3 / 25 + 47 / 69 + 81 = 3 / 72 / 150 = 3 / 73 / 50 = 37350
c/ 36/52/14 + 32/58/74 = 36 + 32 / 52 + 58 / 14 + 74 = 68 / 110 / 88 = 69 / 10 / 88 = 691088
Use the cutting out numbers method to simplify larger numbers.

Learn to determine a digital root

Leave out calculation methods to get interested in the digital root. This notion which name is, I agree, a little bit barbarian and very simple to learn and will serve you in your daily life.

You have already flipped through last pages of a women magazine and you have stopped at the Numerology section. In this section you learn about what the future offers to you, according to your lucky number between 0 and 9.

This number is very easy calculated from your birthday date. Suppose your birthday date is August, 20th 1976 or 08/20/1976. You add then all digits from this date: 2 + 0 + 0 + 8 + 1 + 9 + 7 + 6, that is to say 33. You add now digits from this number 3 + 3 = 6. If you were born on august 20th, 1976, your lucky digit is then number 6.

This digit is called the digital root. It is obtained by successive additions of digits composed of a number to the obtaining of a digit composed between 0 and 9.

Exercises:

Determine the digital root of these numbers:

a/ 54126
b/ 8745
c/ 236514

Answers:
a/ 5+4+1+2+6=18□ 1+8 = 9
b/ 8+7+4+5=24□ 2+4 = 6
c/ 2+3+6+5+1+4=21□ 2+1 = 3

The digital root is a digit between 0 and 9 as a result of the successive addition of summing digits.

Verify an addition result

In the previous chapter, I sold you the digital root as a tool very useful in daily life. In fact, you could use it after each calculation, whatever it is: additions, subtractions, multiplications, or divisions, in order to make sure that the result you get is correct.

I'm going to now show you how to use the digital root in order to verify an addition result.

Illustration:

Previously, we calculated 326521 + 432478 = 758999

Calculate the digital root and its result:
326521: 3+2+6+5+2+1=19□1+9=0□ 1+0 = 1
432478: 4+3+2+4+7+8=28□2+8=10□1+0 = 1
758999: 7+5+8+9+9+9=47□4+7=11□1+1 = 2

The operation we made (326521 + 432478) is an addition.

We then add the digital root of 326521 (which is 1) and that of 432478 (which is 1), that is to say 2.

We verify that the obtained digital root is equal to the digital root of the result 758999 (which is 2).

If we get the same digital root, it means that the calculation is probably correct; if it is not the case, it is certain that the calculation is wrong.

Exercises:

Determine, by the digital root method, if these calculations are wrong:

a/ 359 + 423 = 782
b/ 1026 + 478 = 1504
c/ 586 + 1234 = 1830

Answers:
a/
Determine the digital roots:
359: 3+5+9=17□1+7=8
423: 4+2+3=9
782: 7+8+2=17□1+7=8
Add the digital roots:
8+9=17□1+7=8
The digital roots are the same; the calculation is certainly correct.

b/
Determine the digital roots:

1026: 1+0+2+6=9
478: 4+7+8=19□1+9=1
1504: 1+5+0+4=10□1+0=1
Add the digital roots:
9+1=10□1+0=1
The digital roots are the same; the calculation is certainly correct.

c/
Determine the digital roots:
586: 5+8+6=19□1+9=10□1+0=1
1234: 1+2+3+4=10□1+0=1
1830: 1+8+3+0=12□1+2=3
Add the digital roots:
1+1=2
The digital roots are not the same; the calculation is certainly wrong.

Determine if an addition result is wrong by using the digital roots.

Subtraction techniques

Subtract from left to right; it is simpler

We have said that in mental arithmetic, it is easier to calculate from left to right.

There are many advantages to doing calculations from left to right because we pronounce and write numbers from left to right. However, sometimes, we only need the first important digits, and that would be a waste of time doing all those calculations, as it is the case when we begin from the right.

Illustration:

Suppose that you have this subtraction to solve:

$$
\begin{array}{r}
62 \\
- \quad 47
\end{array}
$$

Mentally, you undertake the calculation from left to right:

6 − 4 = 2

Seeing that, in the next column 2 − 7 will not be possible, take away 1 to the obtained result so 2 − 1 = 1;

Now, in the second column, calculate 12 - 7 instead of 2 – 7: 12 – 7 = 5

Putting results end-to-end, this equals 15.
This subtraction method can be extended to larger numbers:

```
        41268
-       39323
```

Beginning from the left:

4 – 3 = 1, but as the next column calculation is not possible (1 – 9), we take away 1 from the result 1 – 1 = 0 and in the next column, we will calculate 11 – 9 instead of 1 - 9,

In the second column, we calculate 11 – 9 = 2 but as the next column calculation is not possible (2 – 3), we take away 1 from the result 2 – 1 = 1, and in the next column, we will calculate 12 – 3 instead of 2 - 3.

In column 3, we calculate 12 – 3 = 9, and we realize that the next column calculation is possible (6 – 2), so we keep this result (9).

In column 4, we calculate 6 − 2 = 4 and we realize that the next column calculation is possible (8 − 3), so we keep this result (4).

In column 5, we calculate 8 − 3 = 5

We then put all obtained digits end-to-end, getting 01945, which is 1945. We conclude that 41268 − 39323 = 1945.

Exercises:

Solve these subtractions from left to right:

a/ 2568 - 1243
b/ 7236 - 4412
c/ 5214 - 1875

Answers:
a/ 1325
b/ 2824
c/ 3339
Solve your subtractions from left to right:

Subtract each column, beginning from the left, but before writing the answer, look at the next column:
- If the upper is lower than the one in the bottom, write the answer.
- If not, you simplify the digit by 1, write the result, and give the other 1 to the smaller upper number of the next column.
- If the digits are the same, look at the next column to decide how to continue.

Decompose a subtraction with the closed numbers

As in an addition, a subtraction can be simplified from the moment we see numbers containing 0 in the calculation.

The game objective is to modify the subtraction in order to write it in a more legible and easier way to be interpreted by our brain. In fact, it has a natural capacity to see how much things lack in order to fill an empty space.

In this way, 98 is close to 100, but it lacks 2, just as 389 is close to 400 but lacks 11. This observation permits us to calculate in a simpler way.

Illustration:

Suppose that you have this subtraction to solve: 54 - 18.

Mentally, you identify that 18 is close to 20:
18 = 20 - 2

The proposed subtraction can then be written under a simpler form, which is:
54 - 20 + 2 (when taking away 20 instead of 18, we take away 2 more)

That is to say: 34 + 2 = 36

Use the same principle to calculate: 967 - 401 - 198

Mentally, you identify that:
401 is close to 400 (401 = 400 + 1)
198 is close to 200 (198 = 200 - 2)

We can then rewrite the addition under this form:

$967 - (400 + 1) - (200 - 2) = 967 - 400 - 1 - 200 + 2$

Using the regrouping method: $967 - 400 - 200 - 1 + 2$

That is to say: $367 + 1 = 368$

Exercises:

Solve the next subtractions by the decomposing method:

a/ 536 - 122
b/ 846 - 295 - 197

Answers:
a/ $536 - 100 - 22 = 436 - 22 = 414$
b/ $846 - 300 + 5 - 200 + 3 = 346 + 8 = 354$

Use the decomposing method by close numbers in order to simplify subtractions that contain 0.

Cut out numbers in order to solve a subtraction; it is simpler

As in an addition, the mental arithmetic in subtractions by cutting out numbers can considerably simplify the calculation difficulty.

Illustration:

Suppose you have this subtraction to solve:

      541236
-     251012

Mentally, you can undertake the next cutting out:
541236: 54 / 12 / 36
251012: 25 / 10 / 12

And you subtract the cut out numbers:
54 - 25 = 29
12 - 10 = 02
36 - 12 = 24

Putting results end-to-end equals 290224.

Sometimes, calculation seems to be a little more complicated, as in the case when you have to calculate:

```
      845219
-     632587
```

In fact, if you cut out the numbers 2 by 2, 84/52/19 and 63/25/87, the last calculation to solve will be 19 – 87 ...

Likewise, if you cut out the numbers 3 by 3, 845/219 and 632/587, the last calculation to solve will be 219 – 587 ...

The best cutting out is the next one:

84 / 521 / 9

63 / 258 / 7

And you subtract the cut out numbers:

84 - 63 = 21

521 - 258 = 521 – 200 – 58 = 321 – 58 = 263

9 - 7 = 2

That is to say: 212632.

It is then important to determine from the beginning which is the wiser cutting out.

Exercises:

Solve the next subtractions by a wise cutting out method:

a/ 378514 - 227309
b/ 44625 - 21563
c/ 87452 - 63247

Answers:
a/ 37/85/14 − 22/73/09 = 37 - 22 / 85 − 73 / 14 − 09 = 15 / 12 / 05 = 151205
b/ 44/62/5 − 21/56/3 = 44 - 21 / 62 - 56 / 5 − 3 = 23 / 06 / 2 = 23062
c/ 8/74/52 − 6/32/47 = 8 − 6 / 74 − 32 / 52 − 47 = 2 / 42 / 05 = 24205

Use the number cutting out method in order to simplify the large number subtractions.

Subtract numbers to 10, 100, 1000 in the twinkling of an eye

There is an infallible method to solve subtractions from numbers such as 10, 100, 1000, 10000, 100000…. You could challenge your friends, announcing to them that you are capable of giving them the result of 1000000

– 652147 before they can use their calculator. So, how to succeed at this feat?
Illustration:

We are then going to get interested in the next subtraction:

$$1000000$$
$$-\quad\ 652147$$

This technique consists in looking at the bottom number (652147) and determining for each digit its complement to 9 and for the last digit its complement to 10:

What is a complement to 9?
It is the digit to add to another digit in order to get 9.
For example, the 9 complement of 7 is 2 (this is 7+2=9) while the 9 complement of 4 is 5 (this is 4+5=9).
What is a complement to 10?
In the same way, it is the digit to add to another digit in order to get 10.
For example, the 10 complement of 6 is 4 (this is 6+4=10) while the 10 complement of 2 is 8 (this is 2+8=10).

The number to be subtracted, in our example, is 652147

Complement of 6 to 9 □ 3
Complement of 5 to 9 □ 4
Complement of 2 to 9 □ 7
Complement of 1 to 9□ 8
Complement of 4 to 9 □ 5
Complement of 7 to 10□ 3

Putting digits end-to-end equals 347853

So 1000000 − 652147 = 347853.

Important Observations: the bottom number should always have as many digits as 0 in the upper number.

Calculate for example:

```
        10000
-          89
```

The upper number has 4 zeros while the bottom number has 4 digits only. In this calculation, it is necessary to read 2 zeros to the bottom number as bellow:

```
        10000
-        0089
```

The number to be subtracted, in our example, is 0089

Complement of 0 to 9 ☐ 9
Complement of 0 to 9 ☐ 9
Complement of 8 to 9 ☐ 1
Complement of 9 to 10 ☐ 1

Putting digits end-to-end, this equals 9911

That is to say: 10000 − 89 = 9911.

Exercises:

Solve the next subtractions by the based on 10 method:

a/ 100000 - 52478
b/ 10000 - 1563
c/ 100000 - 247
Answers:
a/ 100000 - 52478 = 47522
b/ 10000 − 1563 = 8437
c/ 100000 − 247 = 100000 − 00247 = 99753

When you have to subtract a number of 10, 100, 1000 ... (method based on 10):

Make sure that the bottom number has as many digits as the upper number has zeros,
If this is not the case, re add zeros to the bottom number,
Calculate the complement to 9 of the bottom number digits and the complement to 10 of the last digit of the bottom.
These digits put end-to-end are the subtraction result.

Determine the change in front of the cash register

There is an exercise you could apply to your daily life in order to stimulate your brain to solve mental arithmetic. It is to determine how much the change the storekeeper is going to give you will be when you do the shopping and you pay with a banknote.

Mastering this calculation technique will allow you to make sure that you receive the correct change. Do not forget that good accounts make good friends!

Illustration:

You give a 10€ banknote to pay a 6.53€ purchase.
In order to determine the given change, you solve this calculation:

        10.00
-        6.53

Applying the based on 10 method studied earlier, we get: 3.47€
Now, you give a 20€ banknote to pay your 12.54€ purchase

To determine the given change, you solve this calculation:

        20.00
-       12.54

Related to the based on 10 method, the variant consists of treating differently each number's first digit:

To the first digit of the upper number (2) we subtract the first digit of the bottom number (1) + 1 = 2, that is to say 0

Then, we apply the based on 10 method to the other digits of the bottom number (2, 5 and 4) situated under the upper number zeros, that is to say: 746.

Putting digits end-to-end, we get 07.46 so 7.46€

Now, you give a 50€ banknote to pay your 29.95€ purchase

To determine the given change, you solve this calculation:

        50.00
-       29.95

To the upper number's first digit (5) we subtract the bottom number's first digit (2) + 1 = 3, that is to say 2.

Then, you apply the based on 10 method to the other bottom digits (9, 9 and 5) situated under the upper number zeros, that is to say: 005.

Putting digits end-to-end, we get 20.05 so 20.05€.

Finally, you give a 200€ banknote to pay your 78.32€ purchase
To determine the given change, you solve this calculation:

```
        200.00
-       078.95
```

We realize that the bottom number has fewer digits (4) than the upper number (5). We add then a 0 to the bottom number.

To the upper number's first digit (2) we subtract the bottom number's first digit (0) + 1 = 1, that is to say: 1.

Then, we apply the based on 10 method to the other bottom digits (7, 8, 9 and 5) situated under the upper number zeros, that is to say: 2105.

Putting digits end-to-end, we get 121.05, so 121.05€.

Exercises:

a/ 10 – 3,69
b/ 50 – 25,44
c/ 200 – 24,72

Answers:
a/ 6.31
b/ 24.56
c/ 175.28

To determine the given change, apply the based on 10 method or its variant to multiples of 10, for example, 20, 50 or 200.

Verify a subtraction result

As in an addition, it is possible to use the digital root in order to verify a subtraction result.

Illustration:
Previously, we calculated 845219 - 632587 = 212632
Calculate the digital roots of numbers below:
845219: 8+4+5+2+1+9=29□2+9=11□1+1 = 2
632587: 6+3+2+5+8+7=31□3+1=4
212632: 2+1+2+6+3+2=16□ 1+6=7

The operation we did (845219 - 632587) is a subtraction.

We subtract then the digital root of 845219 (which is 2) and the one of 632587 (which is 4), that is to say -2. If the result is negative, it is necessary to add to it a 9, that is to say -2 + 9 = 7. If the result is positive, it could be used.

We verify that the gotten digital root is the same as the result digital root 212632 (which is 7).

If we get the same digital root, calculation is probably correct; if it is not the case, calculation is certainly wrong.

Exercises:
Determine, by the digital method, if these calculations are wrong:

a/ 658 - 312 = 346
b/ 3627 - 1265 = 2372
c/ 4797 - 524 = 4273

Answers:
a/
Determine digital roots:
658: 6+5+8=19 □ 19=10 □ 1+0=1
312: 3+1+2=6
346: 3+4+6=13 □ 1+3=4
Subtract digital roots:
1-6=-5 □ a negative result, we add9 □ -5+9=4
The digital roots are the same; calculation is correct.

b/
Determine digital roots:
3627: 3+6+2+7=18 □ 1+8=9
1265: 1+2+6+5=14 □ 1+4=5
2372: 2+3+7+2=14 □ 1+4=5

Subtract digital roots:

9-5=4

The digital roots are not the same; calculation is certainly wrong.

c/

Determine digital roots:

4797: 4+7+9+7=27□2+7=9

524: 5+2+4=11□1+1=2

4273: 4+2+7+3=16□1+6=7

Subtract the digital roots:

9-2=7

The digital roots are the same; calculation is correct.

Determine if a subtraction result is wrong by using digital roots.

Multiplication techniques

Multiplication tables - all you need

The first good news in this chapter is that you need only one tool to be able to calculate complex multiplication results. This tool is called "multiplication tables". It is the only indispensable knowledge without which you cannot expect to solve a multiplication. The rest is only techniques and tricks to be taught to you in the next pages.

First of all, I remind you the multiplication b.a.-ba. Take all the time you need, but make sure you know by heart the next tables:

2 x 1 = 2   3 x 1 = 3   4 x 1 = 4   5 x 1 = 5   6 x 1 = 6

2 x 2 = 4   3 x 2 = 6   4 x 2 = 8   5 x 2 = 10   6 x 2 = 12

2 x 3 = 6   3 x 3 = 9   4 x 3 = 12   5 x 3 = 15   6 x 3 = 18

2 x 4 = 8   3 x 4 = 12   4 x 4 = 16   5 x 4 = 20   6 x 4 = 24

2 x 5 = 10   3 x 5 = 15   4 x 5 = 20   5 x 5 = 25   6 x 5 = 30

2 x 6 = 12   3 x 6 = 18   4 x 6 = 24   5 x 6 = 30   6 x 6 = 36

2 x 7 = 14   3 x 7 = 21   4 x 7 = 28   5 x 7 = 35   6 x 7 = 42

2 x 8 = 16   3 x 8 = 24   4 x 8 = 32   5 x 8 = 40   6 x 8 = 48

2 x 9 = 18   3 x 9 = 27   4 x 9 = 36   5 x 9 = 45   6 x 9 = 54

2 x 10 = 20   3 x 10 = 30   4 x 10 = 40   5   x
10 = 50   6 x 10 = 60

7 x 1 = 7   8 x 1 = 8   9 x 1 = 9   10   x
1 = 10
7 x 2 = 14   8 x 2 = 16   9 x 2 = 18   10   x
2 = 20
7 x 3 = 21   8 x 3 = 24   9 x 3 = 27   10   x
3 = 30
7 x 4 = 28   8 x 4 = 32   9 x 4 = 36   10   x
4 = 40
7 x 5 = 35   8 x 5 = 40   9 x 5 = 45   10   x
5 = 50
7 x 6 = 42   8 x 6 = 48   9 x 6 = 54   10   x
6 = 60
7 x 7 = 49   8 x 7 = 56   9 x 7 = 63   10   x
7 = 70
7 x 8 = 56   8 x 8 = 64   9 x 8 = 72   10   x
8 = 80
7 x 9 = 63   8 x 9 = 72   9 x 9 = 81   10   x
9 = 90
7 x 10 = 70   8 x 10 = 80   9 x 10 = 90   10   x
10 = 100

Multiply by 2, 4 and 8 before the calculator

Multiplying by 2 is easy and can be used to quickly solve simple calculations. This way, 26 x 2 would be the same as 26 + 26 is 52.

Thanks to multiplication by 2, we can easily multiply by 4 and by 8. In fact, it is enough to note that:

Multiplying by 4 would be the same as multiplying two times by 2,
Multiplying by 8 would be the same as multiplying three times by 2.

In this way, 24 x 4 = 24 x 2 x 2 = 48 x 2 = 96

And 13 x 8 = 13 x 2 x 2 x 2 = 26 x 2 x 2 = 52 x 2 = 104

Exercises:
Solve these multiplications:

a/ 21 x 4
b/ 17 x 8
c/ 54 x 4

Answer:
a/ 84

b/ 136
c/ 216

Multiplying by 2 is an easy to do operation. Multiplying by 4 would be the same as multiplying two times by 2,
Multiplying by 8 would be the same as multiplying three times by 2.

Multiply by 5, 25 and 50 before the calculator

During the study of addition techniques, we saw that there is a method that allows you to considerably simplify calculations and that consists in making zeros appear in the operation. In the case of a multiplication, it is easy to make zeros appear when we multiply 2 x 5, since the result is 10.

Keeping that in mind, it is very simple to solve multiplications by 5, by 50, and even by 25.

In fact:

In order to multiply by 5, begin with a multiplication by 10, and then divide by 2,

In order to multiply by 50, begin with a multiplication by 100, and then divide by 2,
In order to multiply by 25, begin with a multiplication by 100, and then divide two times by 2.

Illustration:

We need to calculate 18 x 5
Begin calculating 18 x 10 = 180
Then, we calculate 180 / 2 = 90
So 18 x 5 = 90
Calculate 6.2 x 50
Begin calculating 6.2 x 100 = 620
Then, we calculate 620 / 2 = 310

Calculate 246 x 25
Begin calculating 246 x 100 = 24600
Then, we calculate 24600 / 2 = 12300
And then again 12300 / 2 = 6150

Exercises:

Solve these multiplications:

a/ 22 x 5
b/ 16 x 50

c/ 1,4 x 25

Answers:
a/ 110
b/ 800
c/ 35

Multiplying by 5 would be the same as multiplying by 10 and then, divide by 2,
Multiplying by 50 would be the same as multiplying by 100 and then, divide by 2,
Multiplying by 25 would be the same as multiplying by 100 and then, divide two times by 2.

Multiply by 11 faster than anyone

There is a digit that you are going to love when solving multiplications. It is the number 11. Nowadays, many people tremble at the idea of solving, mentally, a multiplication with the digit 11. According to me, it is the type of situation I love. Soon, you are going to be able to challenge your friends by betting that you are capable of solving mentally a multiplication by 11 faster than them...

For the digits between 1 and 9, this is very simple, since the multiplication by 11 consists in doubling the digit we multiply, this way:

1 x 11 = 11 (we double 1)
2 x 11 = 22 (we double 2)
3 x 11 = 33 (we double 3)
And we continue like this up to 9 x 11 = 99

Nevertheless, what does it happen when multiplying by 11 less friendly digits as 0.63 or 321?

Illustration:

Today, you arrive to your office, and you want to buy some stamps. Each stamp has a face value of 0.63€, and you need 11 stamps. How much change should you prepare?

We are going to calculate the price to be paid for these stamps. A stamp costs 0.63€ so, 11 stamps cost 11 x 0.63:

We consider that 0.63€ equals to 63 cents. We calculate then 63 x 11.

In order to solve it, we begin framing 63 with zeros, that is to say:

0630

Next, starting from the right, we add digits 2 by 2, that is to say:

0 + 3 = 3
3 + 6 = 9
6 + 0 = 6

And we put the calculated digits end-to-end, which is 693. That is to say: 63 x 11 = 693.

In order to get the result of 0.63 x 11, we divide the previous result by 100, which is to say 6.93.

Therefore, 0.63€ 11 stamps will cost you 6.93€.

***

Calculate 58 x 11.

We mentally write 0580 and we add digits 2 by 2 starting from the right:

0 + 8 = 8

8 + 5 = 13, we keep the 3 and the 1 will be added to the next calculation.

5 + 0 = 5, to which we add the previous calculation 1, which is 6.

We conclude then that 58 x 11 = 638.

***

That works with even bigger digits.

Calculate 6239 x 11

We mentally write 062390, and we add digits 2 by 2 starting from the right:

0 + 9 = 9

9 + 3 = 12, we keep the 2 and the 1 will be added to the next calculation.

3 + 2 = 5, to which we add the previous calculation 1, which is 6.

2 + 6 = 8

6 + 0 = 6

We conclude then that 6239 x 11 = 68629.

Exercises:

Solve these multiplications:

a/ 22 x 11
b/ 685 x 11
c/ 56 x 11

Answers:
a/ 242
b/ 7535
c/ 616

In order to multiply by 11,
Mentally frame your number with zeros.
Add digits 2 by 2 starting from the right.
Put digits end-to-end in order to form the result.

Multiply from left to right

When solving a mental calculation, multiply from left to right rather than from right to left, as taught at school, to solve a written multiplication. This will allow you to simplify calculation by giving, from the beginning, a good estimation of the calculation result.

Illustration:

We need to calculate 241 x 4

Beginning from the left, we have:

2 x 4 = 8
4 x 4 = 16. We keep the 6, and the 1 is added to the previous calculation (the 8 becomes then 9)
1 x 4 = 4

From the first calculation, we know that the answer will be between 800 and 900. Then, we gradually refine the calculation in order to conclude that 241 x 4 = 964.

***

Calculate 745 x 3

Beginning from the left, we have:

7 x 3 = 21
4 x 3 = 12, we keep the 2 and the 1 is added to the previous calculation (21 becomes then 22).

5 x 3 = 15, we keep the 5 and the 1 is added to the previous calculation (2 becomes then 3).

We conclude that 745 x 3 = 2235.

Exercises:

Solve these multiplications from left to right:

a/ 431 x 3
b/ 124 x 6
c/ 12432 x 2

Answers:
a/ 1293
b/ 744
c/ 24864
Solving a multiplication from left to right, allow, from the beginning, estimating the result magnitude and it can simplify calculation.

Multiply to 3 digits by close numbers

What do these multiplications have in common: 24 x 22, 16 x 12, 56 x 52, 78 x 73, 91 x 95?

All these operations show the particularity to multiply two 2-digit numbers beginning with the same first digit: 24 and 22 begin with 2, 16 and 12 begin with 1, 56 and 52 begin with 5, 78 and 73 begin with 7, finally, 91 and 95 begin with 9.

If you can identify these situations, you will be in a great position in order to mentally solve these operations because there is a calculation technique of close number multiplication.

Illustration:

Calculate 24 x 22

We notice that these 2 numbers are close to 20, in fact:
24 = 20 + 4
22 = 20 + 2

To 24 we add the 2 from 22 = 20 + 2, that is to say 26

(Or to 22 we add the 4 from 24 = 20 + 4 that is to say 26).

We multiply this 26 result by the close number, which is 20:

26 x 20 = 520

We multiply each other the differences to 20 from 24 and 22, which is 4 x 2 = 8.

We add this result to the previous result: 520 + 8 = 528.

We conclude that 24 x 22 = 528

***

Calculate 56 x 52

We notice that these 2 numbers are close to 50, in fact:

56 = 50 + 6

52 = 50 + 2

To 56 we add the 2 from 52 = 50 + 2, that is to say: 58.

(Or to 52 we add the 6 from 56 = 50 + 6, that is to say: 58)

We multiply this result, 58, by the closed number, which is 50:

We saw that in order to multiply by 50, it was easier to multiply by 100 and then divide by 2: 58 x 100 = 5800 and 5800 / 2 = 2900. So 58 x 50 = 2900

We multiply by each other the differences to 50 from 56 and 52, which is 6 x 2 = 12.
We add this result to the previous result: 2900 + 12 = 2912

We conclude that 56 x 52 = 2912

***

We can extend the method to larger numbers; let's calculate 232 x 211.

We notice that these 2 numbers are close to 200, in fact:
232 = 200 + 32
211 = 200 + 11

To 232 we add the 11 from 211 = 200 + 11, that is to say: 243
(Or to 211 we add the 32 from 232 = 200 + 32, that is to say: 243)
We multiply this 243 result by the closest number, which is 200:

We saw that in order to multiply by 200, it was simpler to multiply by 100 and then by 2: 243 x 100 = 24300 and 24300 x 2 = 48600. So, 243 x 200 = 48600.

We multiply by each other the differences to 200 from 232 and 211, which is 32 x 11 = 352 (with the multiplication by 11 method studied before).

We add this result to the previous result: 48600 + 352 = 48952

We conclude that 232 x 211 = 48952.

Exercises:

Solve these multiplications by using the close number method:

a/ 43 x 45
b/ 66 x 64
c/ 332 x 306

Answers:
a/ 1935
b/ 4224
c/ 101592

The multiplication by close number method is possible when we multiply between them the same larger two numbers having the same first digit.

Multiply by decomposing: simplify calculation
When we need to solve a multiplication, the first reflex to adopt is to look for a way to simplify the calculation. We previously saw very powerful methods that simply allow solving complex calculations. This way, multiplying by 2, by 4, by 8, by 5, by 25, by 50 and even by 11 can be done simpler than all other type of calculation.

That is why the decomposing method is going to be attached to the given operation modification in order to make digits appear from which we have a trick to multiply easier.

Illustration:

Face to a given multiplication, the first stage consists in defining which number(s) we are going to create in order to simplify the calculation:

We can try to create a 2, a 4, an 8, a 5, a 25, a 50, an 11, it means, two close numbers.

Calculate for example 32 x 22
We can notice here that 22 = 11 x 2, we can then create two close numbers that are 11 and 2!

So, 32 x 22 = 32 x 11 x 2.
By the previous studied method, 32 x 11 = 352,
And 352 x 2 = 704,
So, 32 x 22 = 704.

***

Calculate 16 x 4.5

We can notice that 16 = 8 x 2.
So, 16 x 4.5 = 8 x 2 x 4.5 = 8 x 9 = 72.

***

Calculate 13 x 150

We can notice here that 150 = 50 x 3. We can then create a friend number, which is 50!

So, 13 x 150 = 13 x 3 x 50 = 39 x 50.
By the previous studied method, multiplying by 50 would be the same as multiplying by 100 and then dividing by 2:
39 x 100 = 3900
and 3900 / 2 = 1950.
So, 13 x 150 = 1950.

***

Calculate 66 x 32

We can notice here that 66 = 33 x 2. We can then create a friend number, which is 2, but mostly we have two close numbers, which are 33 and 32.

So, 66 x 32 = 2 x 33 x 32
By the close number method, we can calculate 33 x 32:

We notice that these 2 numbers are close to 30, in fact:
33 = 30 + 3
32 = 30 + 2

To 33, we add the 2 from 32 = 30 + 2, that is to say: 35.

(Or to 32, we add the 3 from 33 = 30 + 3, which is to say 35)
We multiply this 35 result by the close number, which is 30:
35 x 30 = 1050

We multiply between them the last digits from 33 and 32, that is 3 x 2 = 6.
We add this result to the previous result: 1050 + 6 = 1056.
We conclude that 33 x 32 = 1056.

So, 66 x 32 = 2 x 1056 = 2112.

Exercises:

Solve these multiplications by decomposing numbers:

a/ 66 x 15
b/ 126 x 61
c/ 75 x 52

Answers:
a/ 66x15 = 11x6x15 = 11x90 (the multiplication by 11 method) □ 990
b/ 126x61 = 2x63x61 (the closed number method)□ = 7686

c/ 75x52 = 3x25x52 (the multiplication by 25 method)☐ 3900

In order to simply a multiplication, try to create numbers that allow applying a calculation method:
Multiplication by 2, 4 or 8,
Multiplication by 5, 25 or 50,
Multiplication by 11,
Multiplication by close numbers.

Simplify digits to solve a multiplication

When you do not get to apply the previous technique consisting of creating numbers in the calculation, numbers for which you have a calculation trick, it could be life-saving to try the simplification method.

Illustration:
Let's try to mentally calculate 53 x 89.

It is difficult in this example to decompose the calculation in order to create a 2, 4, 8, 11, 5, 25 or 50.
It is difficult too to create close numbers.

On the contrary, we can note that 89 = 90 − 1 and it is simpler to mentally multiply a number by 90 than by 89.

So, 53 x 89 = 53 x (90 − 1) = 53 x 90 − 53 x 1 = 4770 − 53 = 4717.

***

In the same way, we can calculate 41 x 17.

Noting that 41 = 40 + 1, we can write that 41 x 17 = (40 + 1) x 17 = 40 x 17 + 1 x 17 = 680 + 17 = 697.

Exercises:
Solve these multiplications by simplifying numbers:

a/ 62 x 35
b/ 48 x 15
c/ 23 x 34

Answers:

a/
62x35=(60+2)x35=60x35+2x35=2100+70=2170
b/ 48x15=(50-2)x15=50x15-2x15=750-30=720
c/
23x34=(20+3)x34=20x34+3x34=680+102=782

In order to simplify a multiplication, if the decomposing method fails, we can try to apply the simplification method.

Cut out numbers in order to solve a multiplication

As we previously saw in additions and subtractions, cutting out a multiplication number can be a very efficient method in order to simplify calculations.

This technique is particularly adapted when we multiply a larger number by a short number.

Illustration:

Try to mentally calculate 12331528 x 3.

This type of calculation can be solved by multiplying from left to right, the method that we previously studied. For all that, some multiplication stages in this operation are going to create remainders (Ex. 5 x 3 then 8 x 3) that are going to make the mental calculation more delicate in a larger number.

The recommended method here is to cut out the larger number in shorter ones of only one or two digits. The cutting out will be judicious in order to limit to the maximum multiplications creating remainders.
In our example, we can in this way solve the next cutting out:

12331528 x 3 = 12//33//15//28 x 3 and to solve these multiplications:
12 x 3 = 36
33 x 3 = 99
15 x 3 = 45
28 x 3 = 84

Putting results end-to-end, we get 12331528 x 3 = 36994584.

***

In the same way, we can calculate 211523 x 4.

Applying the cutting out 21//15//23 x 4 = 84//60//92 from which we conclude that 211523 x 4 = 846092.

Exercises:

Solve these multiplications by cutting out numbers:

a/ 19241 x 4
b/ 181217 x 5
c/ 342514 x 3

Answers:
a/ 19241 x 4 = 19//24//1 x 4 = 76//96//4 = 76964
b/ 181217 x 5 = 18//12//17 x 5 = 90//60//85 = 906085
c/ 342514 x 3 = 34//25//14 x 3 = 102//75//42 = 1027542

In order to solve a multiplication implying a large number and a short number, we can wisely cut out the larger one in order to simplify the calculation.

Mental multiplication of very large numbers: it is possible

The previous cutting out method works very well when there is a large number by a short number multiplication, frequently limited to a digit. This tells us that, after having surprised your friends with one of these kinds of calculations, these ones will challenge you to a more complex mental solution.

How will you react when the teasing of them will cause them to ask you to mentally solve this calculation: 121423 x 100002?

Do not worry, applying step by step the method I'm going to teach you, and with a lot of practice, you will soon get to make the achievement of mentally getting this calculation's result!

Illustration:

In order to illustrate the method, we are going to star from a more modest example.

Let's try to mentally give the result of 768 x 997.

Using the close number method in order to say that:

768 is closed to 1000 because 768 = 1000 − 232.

997 is closed to 1000 because 997 = 1000 − 3.

From 768 we subtract the 3 from 997 = 1000 − 3, that is to say: 765.

(Or to 997 we subtract the 232 from 768 = 1000 - 232 that is to say too 765 even it is more complicated to calculate).

We multiply this 765 result by the close number, which is 1000:

765 x 1000 = 765000

We multiply each of the differences to 1000 from 768 and 997, which is 232 x 3 = 696.

We add this result to the previous result: 765000 + 696 = 765696.

We conclude that 768 x 997 = 765696.

Note that it is enough to join the first found result, 765, to the second found result, 696, in order to get the final result 765696.

***

In the same way, we can calculate 121423 x 100002.

Use the close number method, in order to say that:
121423 is closed to 100000 because 121423 = 100000 + 21423
100002 is closed to 100000 because 100002 = 100000 + 2

To 121423 we add the 2 from 100002 = 100000 + 2 that is to say 121425
(Or to 100002 we add the 21423 from 121423 = 100000 + 21423 that is to say 121425, even if it is more complicated to calculate).
We multiply this 121425 result by the close number, which is 100000:
121425 x 100000 = 12142500000.
We multiply each other by the differences to 100000 from 121423 and 100002, which is 21423 x 2 = 42846.
We add this result to the previous result: 12142500000 + 42846 = 12142542846.

We conclude that 121423 x 100002 = 12142542846.

Note that it is enough to join the first found result, 121425, to the second found result, 42846, in order to get the final result 12142542846.

Exercises:

Solve these multiplications by the large number method:

a/ 1123 x 1002
b/ 886 x 998
c/ 8952 x 9995

Answers:
a/ 1125246
b/ 884228
c/ 89475240

The examples below all have a common point. Calculations that they create imply only numbers having the same digits as the other. We have in this way calculated a multiplication with two numbers of 3 digits and then, a multiplication with two numbers of 6 digits. This implies that the used close number is the same for the calculation of two numbers: this way, for 768 x 997 the close number is 1000, for 121423 x 100002 the close number is 100000. All of that makes it interesting to solve multiplications of the type of 9997 x 96, it means, that it implies two numbers that do not have the same quantity of digits.

The method to solve this kind of multiplication is a little bit more complicated, but it is easy assimilated with a little practice.

Illustration:

Let's try to mentally get the result of 9997 x 96.

We can use the close number method in order to say that:
9997 is closed to 10000 because 9997 = 10000 − 3.
96 is close to 100 because 96 = 100 − 4
Let's mentally show the calculation as follows:

Numbers multiplied between them
     Difference related to close number

Numbers		Difference
9997	-	03
96	-	04

We begin with the larger number (9997) to which we subtract the 4 from 96 = 100 – 4.

The subtlety consists in verifying that, in the chart, 96 is lined up with the 99 from 9997. This means that the 4 must be subtracted from the 99 from 9997 and not to the 7 from 9997.

We get, then, the 9597 number.

Next, we multiply each of the differences related to close numbers, which are 03 x 04 = 12

We put end-to-end the calculated results; that is to say: 9997 x 96 = 959712.

Let's try to mentally get the result of 10121 x 1003.

Inspiring us on the close number method in order to say that:

10121 is closed to 10000 because 10121 = 10000 + 121.

1003 is closed to 1000 because 10003 = 1000 + 3.

Let's mentally show the calculation as follows:

Numbers   multiplied   between   them
        Difference related to the close number
10121 + 121
1003   + 003

We begin with the larger number (10121), to which we add the 3 from 1003 = 1000 + 3.
The subtlety consists in verifying that, in the chart, 1003 is lined up to the 1012 from 10121. This means that the 3 must be added to 1012 from 10121 and not to 10121.
We get then the number 10151.
Next, we multiply the differences related to close numbers to each other, which are 121 x 003 = 363
We put end-to-end the calculated results; that is to say: 10121 x 1003 = 10151363.

In order to solve a multiplication implying two large numbers, we can core to the close number method from 1 000, 10 000 or 100 000.
Immediately estimate a multiplication result

Sometimes, you will only need to find the first digit of a result, and the number of zeros that follow it, and not all important digits of the result.

There is a simple method that will allow you to determine in a twinkling of an eye the magnitude of a multiplication result.
Illustration:

Let's try to mentally give the magnitude of 32 x 51

Mentally, it can be represented as:
32 is close to 30,
51 is close to 50

Our calculation will give a close result to this one: 30 x 50 = 1500.

In fact, 32 x 51 = 1632.

***

Let's try to mentally give a magnitude of 496 x 42

Mentally, it can be represented as:

496 is close to 500,
42 is close to 40

Our calculation will give a close result to this one: 500 x 40 = 20000.

In reality, 496 x 42 = 20832.
Exercises:

Give an estimation of the magnitude of the multiplication results:

a/ 1523 x 197
b/ 691 x 812
c/ 10320 x 298

Answers:
a/ 1500 x 200 = 300000
b/ 700 x 800 = 560000
c/ 10000 x 300 = 3000000

Round up the two numbers of a multiplication to the close numbers, allowing to get a magnitude of this operation.
Verify a multiplication result

As in an addition and a subtraction, it is possible to use the digital root in order to verify a multiplication result.
Illustration:

Previously, we calculated 121423 x 100002 = 12142542846.

Let's calculate the digital roots of the above numbers:
121423 = 1+2+1+4+2+3 = 13□ 1+3 = 4.
100002 = 1+0+0+0+0+2 = 3.
12142542846=1+2+1+4+2+5+4+2+8+4+6   =
39□3+9 = 12□1+2 = 3.

The operation we did (121423 x 100002) is a multiplication.

We multiply then the digital root of 121423 (which is 4) and the one of 100002 (which is 3); that is to say 12. The digital root of 12 is 1+2 = 3.

We verify that the digital root is the same as the result digital root 12142542846 (which is 3).

If we get the same digital root, the calculation is probably correct; if it is not the case, it is certain that the calculation is wrong.
Exercises:

Determine, by the digital root, if these calculations are wrong:

a/ 125 x 341 = 42625
b/ 97 x 651 = 63147
c/ 1024 x 422 = 432228

Answers:
a/
Determine the digital roots:
125 : 1+2+5=8
341 : 3+4+1=8
426256 :4+2+6+2+5=19→1+9=10→1+0=1
Multiplication of the digital roots:
8x8=64→6+4=10→1+0=1
The digital roots are the same; calculation is correct.

b/
Determine the digital roots:
97 :9+7=16→1+6=7
651 :6+5+1=12→1+2=3
63147 :6+3+1+4+7=21→2+1=3

Multiplication of the digital roots:

7x3=21→2+1=3

The digital roots are the same; calculation is correct.

c/

Determine the digital roots:

1024 : 1+0+2+4=7

422 : 4+2+2=8

43228 :4+3+2+2+8=19→1+9=10→+0=1

Multiplication of the digital roots:

7x8=56→5+6=11→+1=2

The digital roots are not the same; calculation is certainly wrong.

Determine if a multiplication result is wrong by using the digital roots.

Sum up the multiplication techniques

We saw that the techniques and tricks to mentally solve multiplications are many. At the beginning, you will probably feel yourself lost in choosing the technique that will allow you to easily solve the proposed calculation.

I propose you to look back on the different methods that we have discovered together with its main characteristics:

Multiplication by 2, 4, 8:
This technique allows us to easily multiply a number by 4 and by 8.

Multiplication by 5, 25 and 50:
This technique allows us to easily solve a multiplication in which appears one of these numbers: 5, 25 or 50.

Multiplication by 11:
This technique allows multiplying a number by 11 in the twinkling of an eye.

Multiplication by close numbers:
This technique allows multiplying between them 2 or 3 digit numbers that have the same first digit.

Multiplication by decomposing:
This technique consists in simplifying a multiplication, making appear numbers that allow applying one of the above methods.

Multiplication by simplifying:
In the case of multiplications between of 2-digit numbers, this technique can be applied, in case of difficulty, to use the decomposing method. The simplifying method aims to create in the calculation a number multiple of 10 (10, 20, 30, 40 ...., 90) with which it is easier to mentally solve a multiplication.

Multiplication by cutting out:
This technique is particularly adapted to multiplications between a large number and a one digit number. It aims to cut out the large number by packages of shorter numbers that are easier to multiply.

Multiplication of very large numbers:
This technique is very efficient in order to solve multiplications implying large numbers. It is based on the close number method of 1000, 10000, and 100000.

For the purpose of helping you to apply these techniques, I propose to you, here below, some examples that will allow you to understand the approach to be used in order to determine which method is the most efficient technique to solve the proposed calculation. Practicing regularly with the concrete cases, this process will become automatic, and you will very soon be capable of selecting the more pertinent method, in an intuitive way, by observing the numbers that compose the calculation.

Illustration:

Calculate 532 x 3

3 is not a number with which there is a particular multiplication trick☐ elimination of astute methods.
532 can't be easily decomposed in order to create a friend number or close numbers ☐ elimination of the decomposing method.
532 x 3 is not a large number multiplication☐ elimination of the close number methods.
532 x 3 is a large number multipliedby a one digit number☐ utilization of the cutting out method.

We can write 532 x 3 = 5//3//2 x 3 = 5x3 // 3x3 // 2x3
So, 532 x 3 = 15 // 9 // 6 which is 1596.

***

Calculate 63 x 4

4 is a number with which there is a particular multiplication trick☐ utilization of the multiplication by 4 method.

Multiplying by 4 would be the same as multiplying two times by 2.

So, 63 x 2 = 126 and 126 x 2 = 252
Therefore, 63 x 4 = 252.
Calculate 75 x 6

6 is not a number with which there is a particular multiplication trick ☐ elimination of the astute methods.
75 can't be easily decomposed in order to make appear friend numbers or close numbers☐ utilization of the decomposing method.

It is enough to notice that 75 = 25 x 3.

So, 75 x 6 = 25 x 3 x 6 = 25 x 18.

In order to multiply by 25, we have learned that it is enough to multiply by 100 and then divide two times by 2; that is to say:

18 x 100 = 1800
1800 / 2 = 900
900 / 2 = 450

So, 75 x 6 = 450.

***

Calculate 126 x 62

Both numbers are numbers with which there is a particular multiplication trick□elimination of the astute methods.
126 can be easily decomposed in order to create friendnumbers or close numbers□ utilization of the decomposing method.

It is enough to notice that 126 = 2 x 63.

So, 126 x 62 = 2 x 63 x 62.

We observe that 63 and 62 are close numbers having the same first digit.

We can then use the close number method in order to calculate 63 x 62:

We notice that these 2 numbers are close to 60, in fact:
63 = 60 + 3
62 = 60 + 2

To 63 we add the 2 from 62 = 60 + 2; that is to say: 65.
(Or to 62 we add the 3 from 63 = 60 + 3; that is to say, too: 65)
We multiply this 65 result by the close number, which is 60:
65 x 60 = 3900

We multiply between them the last digits from 63 and 62 which are 3 x 2 = 6.
We add this result to the previous result: 3900 + 6 = 3906.

We conclude that 63 x 62 = 3906.

So, 126 x 62 = 2 x 3906 = 7812.

***

Calculate 123 x 104

Both numbers are numberswith which there is a particular multiplication trick□ elimination of the astute methods.

None of the numbers can be decomposed in order to createfriend numbers or close numbers□ elimination of the decomposing methods.

123 x 104 is a large number multiplication□ utilization of the close number method.

123 is close to 100 because 123 = 100 + 23
104 is close to 100 because 104 = 100 + 4

To 123 we add the 4 from 104 = 100 + 4; that is to say: 127.
(Or to 104 we add the 23 from 123 = 100 + 23; that is to say, too: 127)
We multiply this 127 result by the close number, which is 100:
127 x 100 = 12700.

We multiply each by the differences to 100 from 123 and 104 which is 23 x 4 = 92.
We add this result to the previous result: 12700 + 92 = 12792.

We conclude that 123 x 104 = 12792.

Note that it is enough to join the first found result, 127, to the second one, 92, in order to get the final result 12792.

***

Calculate 998 x 97

Both numbers are numbers with which there is a particular multiplication trick☐ elimination of the astute methods.

None of them can be easily decomposed in order to createfriend numbers or close numbers☐ elimination of the decomposing methods.

998 x 97 is a large number multiplication☐ utilization of the close number method.

998 is close to 1000 because 998 = 1000 − 2.
97 is closed to 100 because 97 = 100 − 3.
Mentally we show the calculation as follow:

Chapter 8: Division Techniques

Division is an operation that will be useful to you in all situations where you will need to share. This operation will be of interest to the child who collects Pokémon cards and who wants to share 26 cards among 4 friends, but also to the mother who will share 75 bonbons among her three kids. According to the quantities to be shared, calculation will be more or less easy. The purpose of this chapter is to allow you to mentally solve most of the divisions you will find in your daily life, but also, if you want to, to solve complex divisions in the twinkling of an eye in order to impress your entourage.

This number can be divided? Divisibility criteria

In my introduction to division, I mentioned two examples in which I'm going to briefly stop.

The first case I mentioned is that one about the kid who collects Pokémon cards. He has

26 doubles that he wants to share with his 4 friends. The question is to determine how many cards he will give to each friend. With 26 cards, he can make four 6-card packages. That is 24 cards, and he will keep 2 cards.

The second situation is the one of the mom who wants to share 75 bonbons between her 3 children. We talk here about determining how many bonbons she can distribute to each kid. The answer is given by the division result of 75 by 3, which is 25. She can then give 25 bonbons to each kid, and she will keep none.

In the first case, we divided 26 by 4, and it left 2.
In the second case, we divided 75 by 3 and it left nothing.

We will say this way that 26 is not divisible by 4 because when we solve the calculation there is one remainder. On the contrary, we will say that 75 is divisible by 3 because when we solve the calculation there are not remainders.

When we want to mentally solve a division, it is very useful to know, before we even start the calculation, if there will be remainders in this division or not. There are very simple methods in order to determine if a number is divisible by 2, 3, 4 ... until 13. This method is called the divisibility criteria.

Divisibility criteria   Example
Divisibility by 2:
A number is divisible by 2 if its last digit is paired (it finishes by 0, 2, 4, 6 or 8)   126 is divisible by 2.

133 is not divisible by 2.
Divisibility by 3:
A number is divisible by 3 if the addition of its digits is a multiple of 3.   131 is not divisible by 3 because 1+3+1=5, and 5 is not a multiple of 3.

531 is divisible by 3 because 5+3+1=9, and 9 is a multiple of 3.
Divisibility by 4:
A number is divisible by 4 if its two last digits are a multiple of 4. 311 is not divisible by 4 because 11 is not a multiple of 4.

624 is divisible by 4 because 24 is a multiple of 4.

Divisibility by 5:

A number is divisible by 5 if its last digit is a 0 or a 5. 234 is not divisible by 5 because its last digit is 4.

990 is divisible by 5 because its last digit is 0.

Divisibility by 6:

A number is divisible by 6 if it is at the same time divisible by 2 and by 3.   741 is not divisible by 6 because it is not divisible by 2 (but it is divisible by 3).

234 is divisible by 6 because it is at the same time divisible by 2 and by 3.

Divisibility by 7:

A number is divisible by 7 if the difference between the number of tens and the double of the unit digit is divisible by 7.   176 is not divisible by 7 because $17 - 2x6 = 17-12 = 5$ is not divisible by 7.

553 is divisible by 7 because $55 - 2x3 = 55-6 = 49$ is divisible by 7.

Divisibility by 8:

A number is divisible by 8 if ht + (u/2) is divisible by 4.

In the number: h is the digit of hundreds, t the digit of tens and u the digit of units.   834 is not divisible by 8 because ht + (u/2) = 83 + (4/2) = 85 is not divisible by 4.

616 is divisible by 8 because ht + (u/2) = 61 + (6/2) = 64 is divisible by 4.

Divisibility by 9:

A number is divisible by 9 if its digit addition is a multiple of 9.   445 is not divisible by 9 because 4+4+5=13 and 13 is not divisible by 9,

756 is divisible by 9 because 7+5+6=18 and 18 is divisible by 9.

Divisibility by 10:

A number is divisible by 10 if its last digit is 0.   849 is not divisible by 10 because its last digit is 9.

320 is divisible by 10 because its last digit is 0.

Divisibility by 11:

A number is divisible by 11 if the difference between the pair digit addition and the impair digit addition is divisible by 11.   1354 is not divisible by 11 because 1+5=6 and 3+4=7 and 7-6=1 is not divisible by 11.

1364 is divisible by 11 because 1+6=7 and 3+4=7 and 7-7=0 is divisible by 11.

Divisibility by 12:

A number is divisible by 12 if it is at the same time divisible by 3 and by 4. 525 is not divisible by 12 because it is not divisible by 4 (but it is divisible by 3).

156 is divisible by 12 because it is at the same time divisible by 3 and by 4.
Divisibility by 13:
A number is divisible by 13 if the addition of the number of tens and the quadruple of the unit digits is divisible by 13. 426 is not divisible by 13 because 42 + 4x6 = 42+24=68 is not divisible by 13.

637 is divisible by 13 because 63 + 4x7 = 63+28=91 is divisible by 13. 91 is divisible by 13 because 9 + 4x1 = 9+4 = 13 is divisible by 13.

Criteria given in the above chart are relatively simple to apply, and they give a quick answer when it is about determining if a number is divisible by another number between 1 and 13. Nevertheless, what about the divisibility of 4913 by 17 or of 3141 by 59?

It is again sufficiently easy to adopt an answer to this question thanks to the method called the zero method:

Illustration:

We need to determine if 4913 is divisible by 17.
The method consists in adding or subtracting numbers that are multiples of 17 to the number 4913 in order to make zeros appear.

For example: 4913 + 17 = 4930.

We subtract then the 0; that is to say 493.

We repeat then the first stage: 493 + 17 = 510. Then we subtract the 0, and it remains then 51.

We notice that 51 = 3 x 17.

So, 4913 is divisible by 17.

Determine if 3141 is divisible by 59.
3141 + 59 = 3200. By subtracting the 0, it remains 32.

32 is not divisible by 59; we conclude that 3141 is not divisible by 59.

Exercises:

Determine by the zero method if these numbers are divisible:

a/ 22222 by 41
b/ 2777 by 23

Answers:
a/ 22222 − 2x41 = 2222 − 82 = 22140□ 2214 − 4x41 = 2214 − 164 = 2050□ 205 − 5x41 = 205 − 205 = 0 So, 22222 is divisible by 41.
b/ 2777 + 23 = 2800□ 28 is not divisible by 23, so 2777 is not divisible by 23.

Divide by 2, 4, and 8 before the calculator

Since division is the reciprocal of the multiplication, the mathematic tricks with the numbers 2, 4, and 8 that we studied in the setting of multiplication can be evidently transposed to division, in this way:

Dividing by 4 would be the same as dividing two times by 2.
Dividing by 8 would be the same as dividing three times by 2.

In this way, 64 / 4 = 64 / 2 / 2 = 32 / 2 = 16

And 152 / 8 = 152 / 2 / 2 / 2 = 76 / 2 / 2 = 38 / 2 = 19.

Exercises:

Solve these divisions:

a/ 92 / 4
b/ 184 / 8
c/ 1296 / 4

Answers:
a/ 23
b/ 23
c/ 324

Dividing by 2 is an easy to solve operation.

Dividing by 4 would be the same as dividing two times by 2,
Dividing by 8 would be the same as dividing three times by 2.

Dividing by 5, 25, and 50 before the calculator

When studying the multiplication techniques, we saw that:

In order to multiply by 5, you begin by multiplying by 10 and then dividing by 2.
In order to multiply by 50, you begin by multiplying by 100 and then dividing by 2.
In order to multiply by 25, you begin by multiplying by 100 and then dividing two times by 2.

Since division is the reciprocal of the multiplication, it is easy to keep the way of dividing by 5, by 50 and by 25; in fact:

In order to divide by 5, you begin by multiplying by 2 and then dividing by 10.
In order to divide by 50, you begin by multiplying by 2 and then dividing by 100.

In order to divide by 25, you begin by multiplying two times by 2 and then dividing by 100.

Illustration:

We need to calculate 2250 / 5
Let's begin by calculating 2250 x 2 = 4500.
Then, we calculate 4500 / 10 = 450.
So, 2250 / 5 = 450

Calculate 7250 / 50
Let's begin by calculating 7250 x 2 = 14500.
Then, we calculate 14500 / 100 = 145.

Calculate 9650 / 25
Let's begin by calculating 9650 x 2 = 19300.
Then, we calculate 19300 x 2 = 38600.
And then again: 38600 / 100 = 386.
Exercises:

Solve these divisions:

a/ 6250 / 5
b/ 11300 / 50
c/ 8650 / 25

Answers:
a/ 1250
b/ 226
c/ 346

Dividing by 5 would be the same as multiplying by 2 and then dividing by 10.
Dividing by 50 would be the same as multiplying by 2 and then dividing by 100.
Dividing by 25 would be the same as multiplying two times by 2 and then dividing by 100.

Divide by 9 before the calculator

In the multiplication chapter, I made you love number 11, showing you how simple it is to multiply a number by 11. This time, we are going to do the same for the division but with number 9.

Illustration:

Calculate 20403 / 9

We put down the first digit starting from the left, which is 2.

We add this digit (2), to the digit on its right (0); that is 2+0=2.

We add this digit (2) to the next digit on its right (4); that is 2+4=6.

We add this digit (6) to the next digit on its right (0); that is 6+0=6.

We add this digit (6) to the next digit on its right (3); that is 6+3=9.

The first 4 digits give the result. The last digit gives the remainder; that is 2266, remainder 9. It means 2267, remainder 0.

We conclude then that 20403 / 9 = 2267.

***

This also works with a larger number. Calculate 124523 / 9

We put down the first digit starting from the left, which is 1.

We add this digit (1), to the digit on its right (2); that is 1+2=3.

We add this digit (3) to the next digit on its right (4); that is 3+4=7.
We add this digit (7) to the next digit on its right (5); that is 7+5=12.
We add this digit (12) to the next digit on its right (2); that is 12+2=14.
We add this digit (14) to the next digit on its right (3); that is 14+3=17.

The first 5 digits give the result; the last digit gives the remainder, that is

1 3 7 (12) (14) remainder (17).

When a number has 2 digits, the first one of these digits is added to the previous number, so we have:

1
3
7 + 1 (1 comes from 12 following 7), which is 8.
2 + 1 (1 comes from 14 following 12), which is 3.
4